THE END OF THE LINE

HOW OVERFISHING IS
CHANGING THE WORLD AND
WHAT WE EAT

THE END OF THE LINE

HOW OVERFISHING IS CHANGING THE WORLD AND WHAT WE EAT

CHARLES CLOVER

EBURY PRESS
LONDON

For Harry and Jack

1 3 5 7 9 10 8 6 4 2

Copyright © Charles Clover 2004

Charles Clover has asserted his right to be identified as the author of this work
in accordance with the Copyright, Designs and Patents Act, 1988.

First published in 2004 by Ebury Press,
Random House, 20 Vauxhall Bridge Road,
London SW1V 2SA
www.randomhouse.co.uk

Random House Australia (Pty) Limited
20 Alfred Street, Milsons Point, Sydney,
New South Wales 2061, Australia

Random House New Zealand Limited
18 Poland Road, Glenfield,
Auckland 10, New Zealand

Random House South Africa (Pty) Limited
Endulini, 5A Jubilee Road,
Parktown 2193, South Africa

The Random House Group Limited Reg. No. 954009

Papers used by Ebury Press are natural, recyclable products
made from wood grown in sustainable forests.

Printed and bound by Clays Ltd, St Ives, PLC

A CIP catalogue record for this book
is available from the British Library

ISBN 0-09-189780-7

CONTENTS

INTRODUCTION –
THE PRICE OF FISH

IMAGINE WHAT people would say if a band of hunters strung a mile of net between two immense all-terrain vehicles and dragged it at speed across the plains of Africa. This fantastical assemblage, like something from a *Mad Max* movie, would scoop up everything in its way: predators, such as lions and cheetahs, lumbering endangered herbivores, such as rhinos and elephants, herds of impala and wildebeest, family groups of warthog and wild dog. Pregnant females would be swept up and carried along, with only the smallest juveniles able to wriggle through the mesh.

Picture how the net is constructed, with a huge metal roller attached to the leading edge. This rolling beam smashes and flattens obstructions, flushing creatures into the approaching filaments. The effect of dragging a huge iron bar across the savannah is to break off every outcrop, uproot every tree, bush and flowering plant, stirring columns of birds into the air. Left behind is a strangely bedraggled landscape resembling a harrowed field. The industrial hunter-gatherers now stop to examine the tangled mess of writhing or dead creatures behind them. There are no markets for about a third of the animals they have caught because they don't taste too good, or because they are simply too small or too squashed. This pile of corpses is dumped on the plain to be consumed by carrion.

This efficient but highly unselective way of killing animals is known as trawling. It is practised the world over every day, from the Barents Sea in the Arctic to the shores of Antarctica, and from the tropical waters of the Indian Ocean and the central Pacific to the temperate waters off Cape Cod. Fishing with nets has been going on for at least 10,000 years – since a time when hunters pursued other humans for food and killed woolly mammoths by driving them off cliffs. Yet because what fishermen do is obscured by distance and the veil of water that covers the Earth, and because fish are cold-blooded rather than cuddly, most people still view what happens at sea differently from what happens on land. We have an outdated image of fishermen as bearded adventurers in the mould of friendly Captain Birds Eye, not as overseers in a slaughterhouse.

Eating fish is fashionable and it is consumed with far less conscience than meat. Even many vegetarians see no irony in eating fish. It has become a kind of dietary talisman for Western consumers. Nutritionists tell us that fish is good for us – the best source of fat-free protein and vitamins – and that the omega-3 fatty acids in oily fish give optimum brain function, reduce the danger of heart attacks and strokes, and delay the onset of arthritis and osteoporosis. Studies even indicate that consuming fish slows down the ageing process and can help us lose weight because a fishy diet switches off our hunger hormone, making us feel satisfied on smaller, more nutritious, amounts of food. Skinny models, whom we are encouraged to admire as the ideal shape for the human form, don't need to smoke to stay skinny, but can be satisfied on bird-like portions. All they have to do is eat fish.

Unfortunately, our love affair with fish is unsustainable. The evidence for this is before our eyes. We have seen what industrial technology did to the great whales, the hunting of which is now subject to a worldwide, but not total, ban. I believe we are about to cross another watershed in public thinking – namely, what industrial techniques, unchecked market forces and lack of conscience are doing to inhabitants of the sea. On land a watershed was reached in farming when sprays, fertilisers, food additives and battery-farming techniques used in the raising of crops and animals led to the collapse of farmers' reputations as custodians of the countryside and guardians of the quality of food we eat. The farmers' image is only slowly being rebuilt, amid much suspicion.

Fish were once seen as renewable resources, creatures that would replenish their stocks forever for our benefit. But around the world there is evidence that numerous types of fish, such as the northern cod, North Sea mackerel, the marbled rock cod of Antarctica and, to a great extent, the west Atlantic bluefin tuna, have been fished out, like the great whales before them, and are not recovering. Reassurance from official sources on both sides of the Atlantic that the seas are being 'managed' scientifically is increasingly hard to believe. Enforcement of the rules that are meant to prevail in the oceans has proved wanting almost everywhere. Even in some of the best-governed democracies, experts admit that overfishing is out of control.

The perception-changing moment for the oceans has arrived. It comes from the realisation that in a single human lifetime we have inflicted a crisis on the oceans greater than any yet caused by pollution. That crisis compares with the destruction of mammoths, bison and whales, the rape of rainforests and the pursuit of bush meat. It is caused by overfishing. As a method of mass destruction, fishing with modern technology is the most destructive activity on Earth. There is no exaggeration in saying that overfishing is changing the world. Just as the deep sea has become the last frontier, its inhabitants a subject of fascination to film-makers, so creatures of the shallow seas, such as sharks and sea horses, are on a slide to extinction. Overfishing threatens to deprive developing countries of food in order to provide delicacies for the tables of rich countries, and looks set to deprive tomorrow's generations of healthy food supplies so that companies can maintain profitability today.

As traditional fish stocks crash and others are found as substitutes, overfishing is altering our diet. It is even altering evolution: the North Sea cod has begun to spawn at an earlier age in response to the pressure of fishing. Overfishing has been, and no doubt will be again, a cause of war and international disputes. It is a force in world trade and international relations, and a corrosive agent in state and community politics.

This book argues that, as a result of overfishing, we are nearing the end of the line for fish stocks and whole ecosystems in the world's oceans, and that it is time we arranged things differently. It takes the form of a journey around the world that I made in several stages and forms the record of many conversations about the problems and potential solutions – many of these as controversial as the problems. It reveals the extent of what is happening in the oceans in our name while satisfying our appetite for fish, and shows the true price of fish that isn't written on the menu.

CHAPTER 1

NAILING THE LIE

WEMBURY, NEAR Plymouth, England. Like a film noir, our story begins with a mutilated corpse on a beach. The body belonged to the sixth common dolphin we had found washed up on a mile of beach owned by the National Trust, which has bought up and protected hundreds of miles of coast in England and Wales, though its writ stops at the high-tide mark. There were fresher dolphin corpses round the rocky headland. The large male we were looking at was the most eloquent in death. The broken, twisted beak showed that the dolphin spent its last minutes tearing frantically at something blocking its way to the surface, almost certainly a net.

Strewn along the beach were bodies in varying states of decay, but all showing the same cause of death. Broken beaks and missing lower jaws and flukes pointed to signs of a frantic struggle as the last ten minutes or so of air in the dolphins' lungs ran out. These animals clearly died unpleasant deaths. Some also had random chunks cut out of them that could not be explained by fox damage after being washed ashore. My companion Joan Edwards, the experienced marine conservation officer of the Wildlife Trusts, observed that these wounds and the missing flukes were probably inflicted as someone hacked the bodies out of a net. Some people found dolphin corpses here with stab wounds, inflicted in a fruitless attempt to make them sink.

As to who was responsible for this carnage, the culprits were clearly fishermen who, like the dolphins, follow the shoals of fish that migrate through the English Channel. Which fishermen were responsible and why was less certain. Trawlers passed us all the time, steaming out from Plymouth harbour to the fishing grounds in the Channel. In the darkest detective stories the evidence is inconclusive and we never find enough to point to who exactly was responsible. All we could do here was look for a motive and try and work back.

It was the fifth year that Joan Edwards and her volunteers had walked these beaches and counted mutilated dolphins washed up on England's south west coast. This year, 2003, there were 123 in six weeks, an increase of 25 per cent on the year before. Experts maintain that the corpses probably represented only one tenth of the dolphins actually killed, since the rest either sank or drifted out to sea. It was unlikely the dolphin population could take this kind of attrition. A video film exists – taken by the Government's fisheries research agency – which showed 30 dolphins trapped in a single haul. I spoke to people who had seen this tape, which would cause an outcry if shown on television, but the agency said its whereabouts were unknown. Dolphins of all species – common, striped, bottle-nosed, Atlantic spotted, Risso's, the lot – are protected under European law. Fishermen who net them by accident must throw them back into the sea, even if they are dead. There is even a treaty that states that fishing *must* be stopped if it is shown that it is taking more than one per cent of a cetacean's population every year. But no one knows how many dolphins there are, in the English Channel, or how many there ought to be, so the treaty can never be invoked. The sad washed-up bodies have struck a chord with the television channels and the newspapers of the South West – Joan Edwards hears anger among the morning dog walkers on the beach.

Local fishermen and conservationists seemed quietly to agree that what was killing the dolphins was pair trawling for sea bass. This involves two powerful trawlers pulling a net up to 1 km (⅔rds of a mile long) between them for hours in the hope of intercepting a shoal of bass returning from their spawning grounds in the

Channel. It is a hit and miss operation. Bass swim at varying depths within the water column, their movements unpredictable, making it hard to find them with a fish-finder, the sonar device that highlights fish as grey clouds on a screen. The trawlers therefore spend much of their time towing water. Dolphins prey on bass, swimming with their shoals, though fishermen say they usually see dolphins only when the net is hauled. The herding of the bass into a tight space makes the dolphins think it is dinner time. They surge into the packed shoals of bass and, in attempting to rise to the surface for air, are trapped by the netting.

Pair-trawling is doubly unpopular locally because outsiders – French, Scottish, Dutch and Irish – are the main practitioners of it, and usually operate in larger vessels than those that putter out of Plymouth and Looe to fish the English Channel.

The dead dolphins make headlines on the local television news, and cause flurries of press releases from animal welfare groups, but what angers local people just as much is the squandering of a great West Country resource, the bass. Animal welfare groups such as the Royal Society for the Prevention of Cruelty to Animals tend not to mention what industrial fishing methods are doing to the fish, only what they are doing to dolphins – arguably an example of 'species-ism', the disrespect for other species about which animal welfare groups often complain. Joan Edwards worries about the bass, though, for the Wildlife Trusts are locally-based. The Trusts know that until the pair trawlers appeared, around the mid-1990s, the bass were actually that rare thing, a conservation success story.

The life cycle of the bass is as miraculous in its way as the Atlantic salmon, which migrates to Greenland to feed, or the eel, which swims all the way to spawn in the Sargasso Sea. Bass spawn in the Channel and the Bay of Biscay in areas only recently discovered by fishermen and scientists. Once the bass eggs hatch, the fry somehow find their way on the current back to estuaries around the coast of Britain, some of them far away from their spawning grounds. Estuaries are the nursery areas where the school bass spend their time until they are 4–7 years old. Until the 1980s, bass were prone to being caught in large numbers by inshore fishermen using fixed nets which were also used to catch sea trout and salmon. Then stocks of all three fish declined. The British Government swiftly banned netting in the estuaries, the crucial nursery areas where bass spend most of their lives. Anglers were encouraged to return individual fish they had caught to the sea. The protection of the bass worked for a time and stocks increased. Anglers, who are important to the west of England's tourist economy, began to notice the improvement. Bass are a near-perfect sporting quarry on rod and line as they run close to the shore in summer, fight hard and grow to over 8 kg (15 lbs) in weight. They were also a welcome, if unexpected, catch for local fishermen trawling for lemon soles or cuttlefish.

It was not long before the resurgent bass (for which the EU has, amazingly, yet to impose a quota, though the British Government limits its own licensed boats to 5 tons a month) began to attract the attention of fishermen with larger boats. Some time in the early 1990s the French trawlers discovered where the bass went to spawn in the Channel. This is when the problems began.

Although bass were never as plentiful as cod and haddock once were, the demand for those caught in Channel waters was reliable and there was money to be made. Bass – *loup de mer* in French – usually attracts a higher price per kilo than any other native British fish – so a relatively small amount of bass will pay for a lot of fuel spent catching it. A large grilled bass that fills your plate will cost you £16 in *Platters* restaurant on Plymouth docks. A strip a few inches across as part of a main course will cost £18.50 in *Fifteen* the chic London restaurant run by Jamie Oliver, the most exciting of British TV's superstar chefs. Recipes for bass are widely found in Oliver's cook books and in his hands, bass became a kind of British national dish when he included a recipe called 'Roasted slashed fillet of sea bass – *à la* Tony Blair' in one of his most popular books. He records that it 'went down a treat' when he cooked it for an Anglo-Italian summit between Blair and the Italian prime minister, Massimo D'Alema, in July 1999.

The choice of bass for a meeting of two prime ministers was another confirmation of the ascendant culinary status of fish. The people who choose menus at summits have to be sensitive to the significance of food. In another era, before Britain's farming industry was brought low by scares about industrial farming methods, and in particular by the epidemic of bovine spongiform encephalopathy, or 'Mad cow' disease, that began in the late 1980s and was linked with a human disease CJD, they might well have chosen beef. However, with a Europe-wide ban on British beef still in force, feeding beef – let alone beef on the bone – to one of Italy's top politicians was unthinkable. So the protocol people chose fish, recognising that it had progressed from being cheap food for the masses to being the premium food of a culture whose confidence in intensive livestock farming had been rocked. Bass has firm flesh and a delicate taste. It was British-caught, satisfyingly expensive, reliably low-fat, healthy and chic – the aspirational dish of the day. And wild fish was untainted by associations with industrial methods.

But that's where they were wrong. For Mr Blair's wild bass – I checked with Jamie Oliver, they were wild – could have come from the English Channel pair-trawl fishery. And the pair-trawl fishery appears not only to be responsible for the slaughter of dolphins but for squandering the bass stocks on which the west of England's inshore fishing industry and tourist trade partly depend. Fishermen report that the bass are less plentiful and the average size is lower. These are classic signs of stock collapse. The anglers complain they are not getting their share. David Rowe of the National Federation of Sea Anglers told me: 'There is no doubt that the slaughter in the offshore fishery affects bass stocks. We've noticed a complete lack of larger fish. That is what keeps anglers going, the expectation of catching larger fish. The International Council for the Exploration of the Sea (ICES), which assesses stock in European waters, has said that catches are still within sustainable limits, but their stock assessments for minor stocks tend to limp a year or two behind reality. ICES has called for the European Union to limit catches to their present level. True to form, the EU has yet to do so.

There is no love lost between local fishermen and the pair-trawlers. Bill Hocking, a Looe-based inshore fisherman, told me: 'It takes a tremendous amount of fish to keep these pair-trawlers afloat, let alone to make money. It's slaughter.' Hocking, an advocate of conservation for far longer than most fishermen, believes in fishing only in ways that leave plenty of fish in the sea.

The plight of the dolphins and the bass in the South West was reported as if it was an isolated problem. But I was to stumble upon something that gave their fate a more sinister aspect and showed that the fate of the bass and the dolphins was linked to a wider disaster.

When I arrived in Plymouth the environmentalists told me that the boats to blame for killing the dolphins and stealing the West Country's fish were big 200ft boats built to catch shoals of mackerel and herring – known collectively as pelagic fish, i.e. those that swim in the upper water of the open sea. Pelagic vessels fish with purse seine nets which are set around the shoals and closed off with drawstrings like purses. Since stocks of pelagic fish remain in relatively good shape, pelagic vessels, based mostly in the big east coast Scottish ports or in Shetland, still pay their way. As a result these new highly equipped ships are emblems of wealth and power within the fishing industry and became convenient scapegoats. Rumour had it that it was rich and callous fishermen in these big pelagic boats, rather than journeymen trawlers fishing out of Plymouth, who were primarily responsible for the pair-trawling that was killing the dolphins and stealing the West Country's fish. As bad guys, they fitted the stereotype, but the rumour turned out to be false. The truth was more poignant.

I found Scottish fishermen in Plymouth harbour, but quite different fishermen – cod and prawn men. Their rusting 80ft trawlers, run with a crew of five, had steamed all the way from the east of Scotland in search of a catch. These were still big boats, but smaller and older than the pelagic vessels I had seen pumping out their fish holds full of herring and mackerel at Peterhead docks. Then the penny dropped about why they were so far south – these were vessels that would otherwise have had to tie up for 15 days a month because of emergency restrictions imposed by European fisheries ministers in a half-hearted attempt to protect the cod.

It was impossible not to feel for them. John Watt, 41, skipper of the *Defiance*, and Ian Duthie, only 28 and skipper of the *Uberous*, were bright lads from Fraserburgh. Owing to cuts imposed in the last, but hardly rigorous, attempt to save the cod, they, along with the rest of east Scotland's fishing industry, were struggling to survive. In desperation, they had turned to bass fishing, and this was the second year the catch had been bad. They showed me the receipts from the market indicating how much they were catching. They had landed 113 kilos (250 lb) of bass for five days' steaming, which they had sold for £670 through Plymouth Trawler Agents. It didn't even cover their fuel bill. Since the beginning of the year they'd laid out £25,000 in fuel and oil, and taken only £6,000 for their fish. The men's wages would have to be funded by their overdrafts. They claimed to have seen – which I think meant killed – a single dolphin in their nets. Watt explained that his vessel was 30 years old, worth little on paper and costly to insure. If he applied for a Government decommissioning scheme (as many did that year) he would be lucky to come out with anything once he'd paid off his overdraft.

The overfishing of the North Sea cod therefore played a part in the deaths of the dolphins washed up on Wembury beach and the collapse of the West Country's bass stocks. The whole European sea was caught in a spiral of decline. The fate of cod in EU waters is an example of history repeating itself. Surely the collapse of the cod on the Grand Banks, off Newfoundland, once the greatest stock of codfish in the world, would have been warning enough to prevent other countries, besides Canada, from going down the same route? Apparently not. It is a conundrum of political systems why so few are capable of learning from others' mistakes. Since 1992 – the same year the Grand Banks fishery was finally closed – scientists from ICES have called with increasing urgency for a drastic reduction in fishing effort in the North Sea. Then, as the black blobs on the chart that denote cod stocks have dwindled from size of a penny to the size of a full stop in ICES autumn assessments, they called for the complete closure of the North Sea cod fishery – without success. In an increasingly desperate attempt to bring fishing to a halt, the secretary general of ICES stated in December

2002 that the amount of spawning cod left in the North Sea was 38,000 tons – the tonnage of a single car ferry.

The annual ministers' meeting in Brussels that year was unusually fraught. The fishing industry had decided that the ban on fishing for cod that the scientists were calling for was unacceptable, as it would mean the death of the whitefish industry in Scotland. Britain's politicians, and those of many countries in Europe, came under pressure to provide more fish to catch, even though biology suggested that this would push stocks ever closer to disaster.

Iain Duncan Smith, leader of the Conservative party, wrote an outrageously inaccurate and ill-informed article in *Fishing News*, throwing doubt on the scientists' view that the cod was on the verge of collapse, though no-one had noticed him take much interest in marine biology until then. Tony Blair, the prime minister, made a much-publicised telephone call to the head of the European Commission asking for British fishermen to be allowed to catch more fish. Apparently this wasn't an issue for demonstrating his environmental credentials. A month later, to mollify those who accused him of playing for fishermen's votes just as irresponsibly as Mr Duncan Smith, Mr Blair asked his Strategy Unit to carry out a review of options for making the UK fishing industry sustainable 'in the medium to long term'.

In the short term, however, Mr Blair had managed to raise the number of days fishermen were allowed at sea from seven to fifteen, while British Government ministers solemnly asserted that the reduction in fishing effort was consistent with scientific advice. This was a lie. In fact, while cuts in fishing effort were made, they were long overdue and poorly enforced. The spiral of decline attrition goes on, the opportunities for recovery are not grasped, and with every turn of the downward spiral there is less chance that nature will bounce back.

The blackened dolphin corpses on Wembury beach are merely a visible sign of a greater and more dispiriting problem, the failure to conserve fish stocks, and therefore nature, in European seas. Although governments still like to give the impression that industrialised fishing is scientifically managed and under control, such claims are looking increasingly threadbare. In most oceans of the world, they do not stand up to sustained examination. Without radically more imaginative conservation measures, life in the seas around Europe and elsewhere will more and more resemble a desert.

Many sensible suggestions for running things better have been proposed, nationally and internationally. What several groups are asking around Plymouth is whether fishing for bass on an industrial scale for negligible returns is really the best use of such a valuable fish. Sea anglers point out, as salmon fishermen did before them, that the value of a rod-caught fish is much greater to the local economy than that of a commercially caught one. Rod-and-line fishermen stay in hotels, hire cars, eat in restaurants and catch relatively few fish. Estimates suggest that 2 million people go sea angling in Britain at least once a year. They spend £1 billion on equipment, travel, food and accommodation – about the same amount of economic activity cause by the commercial catching industry. Surely favouring the anglers and restraining the wasteful industrial boats would be a more sensible idea for the West Country economy?

Five years after it was first asked, the EU, with glacial pace, has finally required gill-net fishermen to fit acoustic devices, 'pingers', to warn harbour porpoises that they are approaching nets. These measures are to be introduced at a leisurely pace – from 2007 in the eastern Channel – and the smallest boats will be exempt. It is also to monitor the bycatch of dolphins through a compulsory observer scheme. The British Government is now trialling the use of escape doors for dolphins on pair trawls. But there is scepticism whether escape doors will work since the same idea was tried for sea lions in New Zealand and Tasmania, with mixed success. The sea lions lost flippers and died.

You might ask what is the point of making nets dolphin-friendly if the bass stocks are as close to collapse as local people believe? A closure of the pair-trawl bass fishery would suit local fishermen and anglers. It would save the bass. It would save the dolphins. Yet the British Government is unwilling or unable to press for closure while

Scottish and French commercial fishermen are running out of things to catch – that is how the august Common Fisheries Policy of the European Union operates. In the strange world of EU politics, the political clout is wielded by the most commercial – which means the most industrial, not necessarily the most economically productive fishermen. So what is called the 'displacement' of fishing effort goes on from cod to bass and from bass to lobster, crabs and every other fish or shellfish that desperate fishermen think they can sell. Absurdities accumulate. The reason that the pair-trawlermen will go on fishing at a loss is simply because they will need to demonstrate a track record of catches in order to claim their share if the EU were to impose a quota in the future.

This is no way to run a common sea. The European Union's inability to manage fish stocks for recovery, its indifference to marine wildlife, its craven electoral fear of fishermen and its inability to agree a solution, such as paying fishermen not to fish so that stocks may recover are all interlinked. They are the result of a political accommodation – an exchange of fishing rights for free trade in Europe – with which no one wants to tamper. No one wants to say no to the fishermen, but no Treasury minister wants to pay them not to fish either, so the system teeters on and the politicians hope secretly the white fish boats will go bust before the cod crashes once and for all.

* * *

Lowestoft, England. January. Six hundred years of enterprise directed at pulling fish from the North Sea in all weathers ends like this. The unrepaired doorways and shabby 1930s office buildings on the seafront tell a story of economic collapse. So, too, does the fish dock of what was once one of England's greatest fishing ports, famous the world over. The grey-brown swell and the driving rain cast a mood of gloom, which is worsened by the gaping crack in the wall and the flapping, broken panel of the company secretary's desk in the offices of Colne Shipping, the last company to run a fleet of beam-trawlers fishing for sole and plaice on the town's traditional fishing grounds off Dogger Bank. Once every schoolchild knew that the Dogger Bank was the North Sea's most productive fishing ground, and that Lowestoft was East Anglia's most productive fishing port. Hugh Sims, company secretary of Colne Shipping, explained how in August 2002 the company made one of its hardest decisions in a 58-year history and tied up its fleet of 40-metre (130 ft) trawlers for the last time because the price of fuel had gone up and the number of fish had gone down.

Inshore boats still go out for a night and land a box or two of cod, rays and dogfish, plying a trade not dissimilar from when Lowestoft was founded as a fishing village in the 14th century. There is a run of sprats, someone has discovered, which the Danish sand eel trawlers with their fine mesh nets haven't vacuumed up yet. But the deep-water fleet has gone. Lowestoft lacks a future, except as a retailing and commercial centre for the Norfolk Broads tourist trade, for there's not much to see in Lowestoft. The terrible sadness that seeps out of the place is reminiscent of a mining town where the coal ran out.

When in 1965 Hugh Sims answered an advertisement for a chartered accountant placed by the Boston Deep Sea Fishing Company, Lowestoft was a thriving port with 120 fishing vessels in the dock. There were five fishing companies, each of which employed its own staff – skipper, crew and onshore workers including fitters, mechanics and ice-men. Only your own staff would get up at a minute's notice at 3am so you could catch the tide. The whole town, from the prostitutes in the dockside bars to the proprietors of prosperous shipping companies, celebrated enterprise and extravagance. Successful skippers wore the port's traditional shore-leave suits in garish colours ranging from red to lime green, when they went out on the town. When one successful skipper bought a Mercedes, the managing director of Colne Fishing, as it was then, said: 'Good. Now the others will want one.'

From its heyday in the 1950s well into the 1980s, Lowestoft was the principle supplier of plaice and sole to Billingsgate, London's fish market. Dover sole was a misnomer. The fish boned at your table in the Savoy Grill was more likely to come from the Dogger Bank via a train from Lowestoft. Competition led the five Lowestoft companies to consolidate and cut staff. They moved to fewer, larger trawlers that could stay two weeks at sea. New techniques of netting were tried. Then in 1984 Colne Fishing moved to the Dutch style of beam-trawling, using powerful 2000-horsepower trawlers dragging nets weighted with heavy beams and 'tickler' chains along the bottom to force the sole, plaice and monkfish up into the net. Beam-trawling was a highly efficient way of killing fish. The only disadvantage seen at the time was that it was heavy on fuel.

Slowly through the 1980s and 1990s, the profits declined. The best skippers saw there was little profit left in the southern North Sea, so they moved away, to the south coast of England or the west of Ireland, where their pay packets were larger because they caught more fish. In the end, Colne had to accept that its vessels were too large and too expensive to run for the amount of fish that were left to be caught.

One might think that the collapse of the last fleet catching flatfish on England's east coast would have caused questions to be asked in Parliament, heads to be shaken and Britain's fisheries minister to be held to account. But after an initial rash of media coverage, the story faded. It was too late and the market had moved on. Buyers had switched to other parts of the British coast to find plaice, sole and monkfish. Consumers, if they noticed anything at all, found there were still plaice and sole in the shops.

There are still fish and chips on the menu at Captain Nemo's, the restaurant on Lowestoft's Claremont pier, which offers cod, haddock, skate, plaice, kippers and tuna bake. Yet the inshore boats bring in only cod, skate and dogfish in small quantities, the rest comes from further and further away. Michael Cole, the Lowestoft fish merchant who supplies Captain Nemo's says he now spends a quarter of his morning sourcing fish on the phone, whereas once he would have bought his stock off the docks at dawn and spent the rest of the day selling it.

Mr Cole says he regrets the loss of Lowestoft's deep-sea fleet every day, for the 20 inshore boats cannot catch turbot, brill, witches (deep water sole) or lemon soles. But he still finds his trade exciting because the world is getting smaller. Mr Cole now calls France, Peterhead, Grimsby, the Channel ports and Milford Haven every day. He has swordfish steaks, red mullet and snapper flown in to Heathrow daily. He can now offer fish from the Seychelles, Sri Lanka, Oman, New Zealand and Australia. A local entrepreneur has begun bringing in cod and haddock from Iceland and the Faroes in containers landed at the docks. The downside, he says, is that all the fish he sells is second-hand and, some of it is not as fresh as he would like, as you depend on the attentiveness of the supplier. Lowestoft, like its rivals Hull and Grimsby, now trades in fish rather than catching it. Mr Cole concedes that, as a consumer, you now have even less of an idea whether the fish stocks your dinner comes from are in good shape, even assuming that would form part of your purchasing decision.

The old world of Lowestoft trawlers and captains courageous exists side by side with the new one of global sourcing in the local branch of the newsagent chain W.H. Smith, where you can buy illustrated histories by a local author, Malcolm R. White, on the old fishing companies of the town. Close by are the racks of glossy magazines containing fish recipes for women conscious of their weight and their health, and opposite is a shelfload of cookery books, prominently featuring Nigella Lawson, whose latest summer cookery book makes mouth-watering use of both swordfish and raw tuna, a taste in fish which has come to Britain from the United States and Japan. Past and present in Lowestoft reflect the reality of fishing: mine out the seam and move on. If Europe depended on its own seas for food, draconian solutions to save the fish stocks and return them to historical levels might have been imposed by now. But we live in a global market so Europe simply exports its demand for fish, with all the destructive power this has had in its own waters, to other oceans of the world. In its defence, it may be said, it is hardly alone in so doing.

CHAPTER 2
FEEDING FRENZY

TSUKIJI MARKET, Tokyo, 4.45am. Top of the food chain. They say that if it swims in the sea, it will end up here in Tsukiji, the biggest fish market in the world. Many central markets have moved out to wholesale parks on the city fringes, but Tokyo's fish market remains lodged in the city's heart. This seems appropriate in a country that eats seafood for breakfast, lunch and dinner, and out of intricate vending machines between mealtimes. Tsukiji is more than a market; it is a national shrine to what one of its wholesalers describes defiantly on its website as the 'inexhaustible sea'. It is a place where a national obsession is indulged and that foreigners visit to be amazed. Part of Tsukiji's fascination is that it is old-fashioned and therefore reveals a lot about Japanese custom. Tourists keen enough to get up in the dark for their fix of Japanese food culture face multiple threats to life and limb. Motorised three-wheeled trolleys, powered by dustbin-like engines with steering wheels, are driven at a frantic pace between auction and stalls. Tractors fitted with giant scoops deposit rock-hard frozen tuna in the back of trucks at head height. Normally polite Japanese give no quarter. The best advice is to get out of the way.

The market is set out like this. Two large areas in this single-storey sale-room of the sea are laid out with tuna. A larger wet-fish area sells virtually everything else from salmon to horse mackerel. Peeping out of the few iced polystyrene boxes that are open for show – in a market obsessed with quality everything is kept at minimum temperature – are yellowtail, squid of all sizes, anchovies, snapper, barracuda, butterfish and eye-catching scarlet alfonsinos.

On the huge number of small stalls over the covered road from the main auction rooms you will find anything you might want to consume and many things you might wish to avoid. Finger-long eels from China squirm in buckets. Live amberjack and Pacific cod from farms around Japan wait in tanks to be slaughtered. The notorious fugu, or puffer fish, swims listlessly, waiting to kill some unfortunate diners with the poisons in its skin or liver if it is not properly prepared. Unfeasibly large, phallic barnacles and fresh mussels await specialist buyers. A whole stall of pink, cooked octopus is neatly set out, suckers skywards, tentacles tucked underneath, looking like sea urchins without their spines. Here sits a speciality I have yet to see on any menu: a tray of garfish, their long beaks tucked within the coils of their bodies. The buyers here delight in variety and seasonality. Just in, and attracting good prices, is Pacific saury, a thin, silvery fish shaped a bit like a miniature barracuda.

Here in the most crowded part of the market, where there is a danger of being hit by a shopping cyclist, marine delicacies air-freighted from different oceans, involving incalculable emissions of carbon, are packed into every inch of space. Stalls specialise: one has pink prawns from Mexico, black tiger prawns from Vietnam and Malaysia, another scallops, lobsters and crabs from the north Atlantic, another farmed salmon and fresh salmon eggs from Chile and Norway. The stalls also have a role in the most lucrative business of the market – providing raw fish to a city of 15 million people obsessed with the taste of sashimi, the standard version of Japanese raw fish, and the added design and extravagance of sushi, raw fish in its exquisitely presented haute cuisine form. Top-quality raw fish is everywhere in Tokyo: it just arrives on the side when you order a beer. The stalls here are where the giants – the tuna and swordfish – will be dismembered and jointed into more affordable chunks after the early morning frenzy of the auction is over.

From around 4am the wholesalers arrive to inspect the tuna laid

out on palettes by the auctioneers' staff. Nearly a thousand fresh
tuna, northern bluefin, southern bluefin and bigeye, will be sold in
the next two hours. About the same number, blast-frozen at sea at
–50°C, lie frosted white and gill-less on a floor in an adjoining hall.
A mist rises around the buyers' white rubber boots as the tuna
warm. The most valuable fresh bluefins are slashed along the lateral
line so that the deep red colour, the sign of peak freshness, is visible.
Each fish is labelled with its weight and surmounted by a tiny
sample of the most succulent flesh laid out on a piece of paper for
the buyers to taste. This courtesy is largely disdained by the
wholesalers, who rely on experience, a torch and a tuna hook to
examine each carcass for signs of *yake* or 'meat burning' – the first
sign of decay – which lowers the price.

It's a big auction today because it is the Friday before a public
holiday, explains my guide Hide (pronounced 'hee-day'), a student
in his mid-twenties. He is wearing a fleece emblazoned with the
badge of the Australian tuna company he is to work for in the new
year. The auctioneer rings a bell at 5.30am precisely. The wholesalers, with cap badges denoting their licence to bid, surge into the
stands and the auctioneer starts the sale. Meanwhile, another four
auctions by other firms are being held on other areas of the floor.
Each auctioneer has his own patter, his own particular tone and
vocabulary. To a non-Japanese speaker the auctioneer we are
watching seems to be uttering a particularly frenzied torrent of
howls and yelps. Each price is made in less than ten seconds, the
buyers holding up fingers to indicate what they are prepared to pay,
in an unspoken contract based on trust.

The bluefin represents the peak of evolution in fish – an extraordinary creature that can swim at more than 80 kilometers an hour
(50 mph), accelerate faster than a Porsche, and cross the Atlantic in
its annual migrations. The secret of its ability to turn on such
dazzling bursts of energy is that it is *warm blooded*. Here, however,
what matters is taste and the bluefin is commonly agreed –
unfortunately for it – to be the fish equivalent of Aberdeen Angus
raised on Argentine pampas.

The largest fresh bluefin, and therefore the first to be auctioned, is a giant of 232 kg (510 lb) caught by fishermen with mixed motives off Cape Cod. The New England fishery is allegedly conducted for sport, but money is clearly another motive. As Carl Safina has ably demonstrated in his book *Song for the Blue Ocean*, greed and political influence have kept a fishery going that by rights should have closed or been drastically reduced years ago. Every year since the 1960s the USA's eastern seaboard fishery has taken more bluefins from the threatened west Atlantic population than scientists say is right. The headless carcass on sale in Tokyo was air-freighted from Boston, packed respectfully in plastic and refrigerated within a huge cardboard sarcophagus filled with ice. Tuna fishermen in the United States claim, probably rightly, that the western Atlantic bluefin tuna fishery is under the strictest controls ever. What almost everyone now acknowledges is that the most urgent problem is with the bigger, largely separate population of bluefin tuna that forages in the east Atlantic and migrates into the Mediterranean.

First, a surprise. The Boston fish makes only 4,000 yen per kilo (£9 per pound), or £4,620 for the whole fish. This might seem a lot of money for a fish, but exceptionally large, fresh bluefins can make £50,000 apiece. Hide's expectations were higher. A new trend is exerting downward pressure on the market today. Top-quality bluefin tuna usually fetch over 5,000 yen (£25) per kilo at Tsukiji. Today, even with strong demand, we see the price of sparkling fresh bluefin slump at under 4,000 yen (£20) a kilo. There is a brief surge to 4,600 yen (£23) for one particularly fresh specimen of southern bluefin, caught off Australia and fattened in a 'farm' before being shot through the head. Hide says tuna fattening – he calls it 'farming' – is what has caused the slide in the market. Ironically enough, a fish regarded as over-exploited in every ocean where it swims is in oversupply. Too many fish are in the market, too few in the ocean.

Tuna farming is the only kind of farming I know in which you reap but you don't sow. Tuna fattening, as it should really be called, started in Australia, where it was found that small, overfished southern bluefins could be purse-seined in shoals and released into cages at sea. The fish are fattened on low-value wild-caught fish until the level of oil in their flesh has improved – fattened fish have better oil levels than wild fish – and the price is right. No actual breeding takes place, though Japanese research has shown it to be theoretically possible. Breeding is just uneconomic compared with catching wild fish, as there is no charge or penalty to reflect the fact that the wild fish are over-exploited. Farming could be a good idea if it genuinely meant breeding, but first one would have to ensure the wild fish, the brood stock, was properly looked after. Australia claims to have its southern bluefin fishery under control, but this cannot be said of Europe.

In under a decade, tuna fattening has swept the Mediterranean, where the bluefin tuna has been prized since the ancient Greeks gloried in its slaughter and it was used to feed Roman legions in preparation for battle. The eastern Atlantic population of bluefins is larger than the one remaining in the west. It migrates in a great ellipse out into the Atlantic and ranges into the furthest corners of the Mediterranean. At one time, most forms of tuna-catching were either inefficient or designed to catch only large tuna, like the Sicilian *mattanza*, where tuna are herded into and are gaffed bloodily in an intricate network of set nets. Now French and Spanish purse-seining fleets with the finest fish-spotting and catching equipment are being used to encircle bluefins of all sizes. Once encircled, and before the net has been drawn completely tight around them, the bluefins are transferred to cages. The cages are then slowly towed from where they are caught to their final destination at less than two knots, the slow speed essential to ensure the tuna will not be caught in the nets and drown. Farms have grown up conveniently positioned around the Mediterranean, in Spain, Malta, Sicily, Cyprus, Libya and Turkey, so that the bluefins may reach the coastal holding pens with minimal casualties. The trade has generated a vast amount of business, not only with Japan and for the airlines who carry the tuna there, but also for makers of nets and sea cages.

Tuna farming has killed any hope of monitoring catches accurately. Bluefin tuna is the only fish in the Mediterranean subject to quotas – the pitiful size of most other fish reflects the fact that there are no fisheries controls, not even 200 mile limits, in this crucible of European civilisation. What makes bluefin catches particularly hard to quantify is that the farmed fish are not landed immediately, and that the fish are towed to another part of the sea from where they were first caught.

In the past two years the practice of farming or fattening has migrated from the Mediterranean to the Canaries and westwards to Baja California, where tropical species of tuna are now being fattened for the fusion restaurants of Los Angeles, New York and Toronto, and, of course, for the Japanese market itself.

The numbers of bluefin tuna arriving from Mediterranean and Australian 'farms' can be seen from the number of paper stickers on the carcasses in Tsukiji. This year is the first in which there appears to be an obvious oversupply, a cause of concern in Japan because it has begun to affect the price. 'There were 30,000 tons of bluefin tuna from the Mediterranean this year. It was very good catching in the Med, but in two years maybe there could be a supply problem,' says Hide.

What Hide describes euphemistically as a supply problem means that the breeding stock of bluefins is being slaughtered and soon there will be no adult tuna left to catch. The problem has already occurred to ICES, based in Copenhagen, whose scientists believe that 30,000 tons is roughly double the sustainable catch of tuna. The trouble is that managing bluefin tuna is not really the business of ICES. The management of tuna in the Atlantic is the job of the International Commission for the Conservation of Atlantic Tunas (ICCAT), based in Madrid. Scientists from ICCAT – fondly known to conservationists as the International Conspiracy to Catch All Tunas – assess bluefin stocks and issue management recommendations. And politicians from states that are members of the Atlantic Tuna Treaty – including the USA, Japan and the EU – duly ignore them.

The scientists' latest assessment says that things are in an even worse mess than usual. The western bluefin stock, which runs up the eastern coast of the USA and Canada, has been fished down to a level about a tenth of the size it was in the 1960s. (Just to confuse matters, some tunas are now known to intermingle, crossing the Atlantic to do so, but taking account of this would make managing the stock even more difficult politically, so ICCAT doesn't bother.) In the eastern Atlantic and the Mediterranean, tuna farming has led to the collapse of the system for gathering information about catches and imposing limits. ICCAT says that catches of bluefin tuna from the Mediterranean and eastern Atlantic are currently 32,754 tons. The level at which ICCAT thinks the present population can be sustained is somewhere below 26,000 tons, a much higher figure than ICES said, you will note, but still substantially below present catches. It is therefore virtually guaranteed that the present level of catches will further reduce the breeding population.

The scientists say they are 'strongly concerned' – which isn't quite English, but nor are they – about the failure of catching countries to provide adequate catch returns. They mean principally France. When it comes to the management of fisheries in the Mediterranean, scientific reports read like counsels of despair. The scientists point out that a minimum size for tuna has been in force for the Atlantic and Mediterranean since 1975, and has been ignored ever since. So has a ban on fishing with purse-seine nets from mid-July to mid-August, and a ban on the use of spotter aircraft or helicopters to spot tuna shoals in the month of June. The scientists question, helplessly, whether these measures are or could be enforced. They point out to the governments, who are supposed to ensure ICCAT's rulings are followed, that too many small fish are being caught and that the abrupt increase in large fish catches since 1994 'is of grave concern'.

The explosion of tuna farms led to a wildly unsustainable catch in excess of 50,000 tons in 1996, and nearly as much in the following years, mostly small fish that had no chance to spawn. A cynic would say that catch returns are such a potential source of embarrassment to fishing nations that it makes life easier not to release them at all.

Despite the scientists' grave warnings, the kind of politicians that are sent to ICCAT meetings voted in 2002 to *increase* the quota for the next four years from 29,000 to 33,000 tons. They described this as a multi-annual management plan. I just didn't understand this as I re-read the tuna assessment on the plane home. Had I missed something that would actually conserve the tuna, I asked Sergi Tudela of Spain's Worldwide Fund for Nature (WWF). No, I hadn't missed anything, he said. This was ICCAT. 'This is a gold rush. Everybody is trying to get involved and nobody is seriously trying to achieve sustainable exploitation of bluefin tuna.'

So what one sees on the auction floor at Tsukiji is a genuine scandal. Tsukiji may be one of the few fish markets in the world that does not smell of fish – the Japanese obsession with cleanliness extends everywhere, even to the outer alleyways and gutters, which harbour a stench in most markets around the world but here are hosed down and scrubbed several times a day. But it still stinks, for its daily trade is pushing bluefin tuna daily closer to extinction as surely as the black rhino or the elephant was brought to the brink in Africa's grisly 1980s' epidemic of poaching. The bluefin is a textbook case of unsustainable, unregulated global trade – the sort of thing most people mean when they talk about globalisation.

In fact, there is a perfectly sound global solution – the Convention on Trade in Endangered Species (CITES). Listing the elephant under Appendix 1 of the convention, which means a total ban on trade, almost certainly saved it from being wiped out by unsustainable demand for ivory from, you guessed, Japan. An attempt to list the bluefin tuna was made by well-meaning, conscience-stricken Sweden – a country that has lost its bluefin run altogether – in 1992. Sweden proposed a listing under Appendix 2 of the convention, which permits trade, provided it is strictly controlled at both ends and each traded animal carries document-ation to say it is legally and sustainably caught. Such a listing has since been given to several freshwater fish, including the sturgeons of the Caspian Sea, so it is clearly workable. The 1992 proposal was withdrawn after intense diplomatic pressure from Japan and internal political pressure from tuna lobbyists within the United States. There is no record of what the Japanese did to the Swedes to bring about this withdrawal, whether it was threatening not to sell them Toyotas or Sony TVs, or perhaps erecting more tariff barriers to Volvos. But it worked and the Swedes accepted the inevitable. Realistically, for pressure to build again at CITES there needs to be a stock assessment that shows the bluefin is even more at risk than the Atlantic northern bluefin, categorised as 'endangered' in the Red List compiled by the World Conservation Union. Currently the global stock is listed 'data deficient'. Call me cynical, but I wonder if this could be why Mediterranean countries have been so bad at filing their catch returns?

The tuna farming free-for-all means that the bluefin will be driven either to extinction or into the same remnant population as the blue whale (c. 1,500 individuals). Yet if it is, the finger of blame cannot be pointed just at Japan. Of course, Japan must accept its fair share because it is virtually the only market for bluefin tuna, and its long-line fleet was entirely responsible for fishing out the tuna population off Brazil in the 1970s and 1980s. On the other hand, Japan points out, only a little disingenuously, that it is bound by the rules on free trade. Arguably, it has also done more than any other nation to buy out illegal long-liners operating mostly out of Taiwan. It has so far bought out 140 Taiwanese long-line vessels catching tropical tuna – yellowfin, skipjack and bigeye – which were fishing in competition with its own tropical long-line fleet. It has yet to do much for the temperate-water bluefin.

When it comes to the collapse of the Atlantic giants, the blame for wiping out shoals that could have fed future generations will have to be spread more widely than Japan. The blame will reside in the 75 catching countries, principally in the United States and Europe, which are responsible for managing the largest fisheries. In most terrestrial matters, these claim to be leaders in conservation. In the case of Europe, which has access to the largest remaining population of Atlantic bluefin, cynicism is endemic. Franz Fischler, the European Fisheries commissioner, was asked whether the new tuna farms were eligible for aquaculture subsidies. He replied that under the EU's rules, they were. I have found out since that Spain alone has given 6.5 million Euros (£4.4 million) in subsidies for tuna farming, since 1994. Much, if not all, is likely to have come from EU coffers. An EU official attempted to excuse such subsidies by saying that marine salmon farms also raised fish they didn't breed – these were raised in lakes and rivers in fresh water. I pointed out that these fish were still bred in captivity. So she tried another tack: 'This is something that we are concerned about, but it affects the whole Mediterranean, not just the EU.' So tuna farming goes into the too-difficult box. In any case, the official added with the customary pragmatism of the fisheries directorate, 'This is something we would want to regulate, not close down altogether.'

If the bluefin is squandered, and that time isn't far off, it won't just be governments who are to blame – consumers must shoulder responsibility too. It is not just the Japanese who eat Japanese food. The taste for sushi, already established throughout the East, is spreading into restaurants and supermarkets in other continents far from Japan.

A Western myth about Japan is that bluefin tuna is impossibly expensive and eaten only by the very rich. If it were, the trade might be easier to regulate. The reality is that the £30,000 bluefin is a rarity at Tsukiji. Most bluefins will be consumed by hundreds of people in tiny amounts. The outer perimeter of Tsukiji is circled by restaurants, which Hide and I reached at 9.00am, longing for breakfast after being up since the early hours. We fell into the first restaurant we came to. Hide chose cod and cabbage soup, which sells at about 600 yen (£3). The menu also listed a bluefin sashimi 'set' for 1,100 yen (£5.50). I ordered it on a dual impulse of hunger and curiosity. It came with an elegant garnish of twirled spring onion and cucumber, some wasabi (Japanese horseradish), soy sauce and a pile of dark red flesh with a sensuous texture and a rich taste, weighing in total around 150 g (5 oz). On the side was a box of rice and a bowl of miso soup. The taste was delicious and has remained memorable, as has the guilt that struck me once I had eaten it. That, I resolved, was enough bluefin for a lifetime, or until bluefin numbers miraculously increase. If that is to happen, the Japanese – and everyone else – are going to have to get used to eating fewer fish.

*   *   *

Vigo, Spain, 7.00am. Vigo claims to be the largest port in the world dealing in landings for human consumption. A bigger quantity of smaller fish is landed for processing into fish-meal in the North Sea and Peru, but Vigo sees the spectacular stuff, the big fish, the remaining giants of the deep. Swordfish and sharks entirely covered the floor of the brand-new market building for the Spanish distant-water fleet – a building I was told was constructed with 75 per cent subsidies from the European Union. There are two other enormous halls on the dockside – one for ground (bottom-dwelling) fish, such as cod and hake, and one for pelagic (shallow-swimming) fish, such as anchovy and sardine. But today the showpiece was the distant-water hall because of the catch laid out there, the result of less than two weeks' fishing by just *two* vessels.

The entire floor, save the walkways, was being covered with swordfish ranging in weight 44–241 kg (96–530 lb). As they came off the boat named *Portuguese Sea*, a 28-metre (92-ft) long-liner fishing off the Azores, they were weighed individually on the auctioneer's scales and the weights recorded in the skipper's logbook. Carlos Pacheco, the second skipper of this Portuguese boat, told us as he recorded the catch: 'This is a beautiful life. But we catch some big storms. The last time I thought I was standing on top of a glass of beer. You just couldn't see the deck for foam.'

Swordfish, among the few fish that, when hooked, have been known to attack their captors's boats and sink them, were laid out in rows on a series of pallets. At the end of each line was dumped a pile of sharks. These were mostly fox sharks, distinguished by a long thresher ribbon on top of the tail, and were of low value. But here and there was a blue shark, the flesh of which is more prized. These sharks would probably not have been landed a year ago. Their fins would have been cut off and the rest of the living shark tipped back into the sea. The European Union has decided that soon the whole shark will have to be landed, so either the skippers were paying early respect to that ruling or they thought shark prices were looking up. The haul, of about 20 tons, also included two exotics – a lemon dorado and a moonfish. The swordfish fetch €6 (£4) a kilo, the more marketable sharks €1.5 (£1), the rest 75–80 cents (50–53 pence). Looking down the room and reflecting that this impressive haul was the result of two long-line boats fishing for ten days, one began to have fears for the productivity of the sea.

Raul Garcia, my contact from WWF Spain and a Galician himself, found the sight of 4-metre (13-ft) swordfish reassuring. What we were seeing, he told me, was a conservation success, though this was likely to be short-lived. ICCAT, which manages catches of swordfish as well as tuna, actually succeeded in pushing through a reduction in swordfish quotas for a few years in the late 1990s, prompted by an alarming slump in North Atlantic sword-fish catches and a drop in the size of the fish. Trade sanctions were brought by nations that fished legally against countries that provide flags of convenience for vessels that fish illegally and without reporting their catches. Countries such as Belize duly kicked the pirates off their registers. A boycott on eating swordfish in the United States would appear to have helped. The North Atlantic swordfish stock recovered from the desperate trouble it was in a decade ago. ICCAT has just granted a 10 per cent increase in quota for swordfish, almost certainly too early. Whether this will simply spur another bout of overfishing remains to be seen. Most worrying is that European Union vessels are shortly to be allowed within 160 km (100 miles) of the waters around the Azores, whose 320-km (200-mile) limit was previously open only to its own fishermen and those from Portugal. This will open up the relatively lightly fished waters where the swordfish in Vigo market came from, one of their last refuges, to the rest of the EU long-line fleet. Yet again, the European Union is at the leading edge of an unsustainable trend.

Everywhere else in the ocean there are worrying signs. The swordfish is in a worse state in the southern Atlantic. The state of swordfish in the Mediterranean is desperate, with catches based wholly on small fish, many of which have not had the chance to spawn. Records of catches are atrociously kept. The improvement in the north Atlantic shows what can be done. It bucks the trend, but for how long?

The trend was set out chillingly in an article in the science journal *Nature*. The authors took ten years to evaluate all the major fisheries in the world and found that only 10 per cent of the stocks of large fish – tuna, swordfish, marlin and large ground fish, such as cod, halibut, skate and flounder – present in 1950 were left in the ocean. Ransom Myers, a fisheries biologist based at Dalhousie University in Canada, and co-author of the article, said at the time:

> From giant blue marlin to giant bluefin tuna, and from tropical groupers to Antarctic cod, industrial fishing has scoured the global ocean. There is no blue frontier left. Since 1950, with the onset of industrialised fisheries, we have rapidly reduced the resource base to less than 10 per cent – not just in some areas, not just for some stocks, but for entire communities of these large fish species from the tropics to the poles.

Most worryingly of all, perhaps, the study showed that industrial fisheries take only 10–15 years to reduce any new fish community they encounter to one tenth of what it was before. Myers's colleague Boris Worm, of the University of Kiel in Germany, said:

> The impact we have had on ecosystems has been vastly under-estimated. These are the megafauna, the big predators of the sea and the species we most value. Their depletion not only threatens the future of these fish and the fishers that depend on them, it could also bring about a complete reorganisation of ocean ecosystems, with unknown global consequences.

Myers and Worm found no room for optimism. What they discovered applied not only to the seas of continental shelves but also to areas of open ocean, where people had presumed there were still untapped reservoirs of large fish. Using data from the Japanese long-line fleet, the most widespread operation, which fishes in all oceans except the circumpolar seas, they found that where long lines used to catch ten fish per 100 hooks, they were now lucky to catch one. Daniel Pauly of the University of British Columbia put it this way: 'It is like a hole burning through paper. As the hole expands, the edge is where the fisheries concentrate until there is nowhere left to go.'

The solution suggested by Myers and Worm is simple but drastic: they call for the number of fished killed every year to be reduced by 50 per cent in the populations most at risk. The question of how to make this solution politically acceptable is one of the world's great political problems. Myers says: 'If stocks were restored to higher abundance, we could get just as much fish out of the ocean by putting in only a third or a tenth of the effort. It would be difficult for fishermen initially – but they will see the gains in the long run.' One of the ways to protect the sea's big fish is to create a network of sufficiently large areas where no fishing happens at all: they would act as reserves.

Myers says this might sound drastic, but the alternative is a world in which tuna, sharks and swordfish are merely memories.

We are in massive denial and continue to bicker over the last shrinking numbers of survivors, employing satellites and sensors to catch the last fish left. We have to understand how close to extinction some of these populations really are. And we must act now before they reach the point of no return. I want there to be hammerhead sharks and bluefin tuna around when my five-year-old son grows up. If present fishing levels persist, these great fish *will* go the way of the dinosaurs.

Myers and Worm believe they have explained the magnitude of the change that has gone on within a single human lifetime and provided the 'missing baseline' needed to restore fisheries and marine ecosystems to healthy levels. But as one might expect, many working fisheries scientists and officials, who found themselves accused of presiding over the last 10 per cent of a fishery, were reluctant to accept the results. 'We found there was acceptance of the overall pattern of rapid depletion of communities, but there was more controversy when it came to the current status of individual species, particularly with tuna. Understandably, some fisheries managers find it very hard to accept,' says Myers. The editors of

Nature didn't have the same difficulty. Myers and Worm's conclusions survived both the carping of professional fisheries 'managers' and the magazine's rigorous process of independent peer review. It ended up as the cover story.

As the big fish are fished out, a process of substitution goes on. Big fish are replaced on the menu by smaller fish that are, for a while, more plentiful because the bigger ones have been removed. Or, if they can still be obtained, fish from overseas replace big fish from home waters. This process goes on from Tokyo to Lowestoft and from Billingsgate to Mercamadrid, the largest fish market in Europe. There one trader told me, 'Twenty years ago, the market was not as big. It has grown very quickly. Before the fish came from Chile, Argentina and Spain. Now they come from all over the world.' Ransom Myers, Boris Worm and Daniel Pauly have all described this phenomenon of substitution, which continues to provide the impression of plentiful fish supplies. Daniel Pauly has coined the best names for this process, which has been going on since the 1950s: 'fishing down food webs'. Some call it 'fishing down the food chain'. I prefer webs, as they have more connections than a chain. In Pauly's view, all we will eventually be left with is jellyfish and plankton. And if you think we will then stop fishing, Pauly has news for you. He reports that one Georgia fisherman is already making a living by sending 22,500 kg (50,000 lb) of jellyfish a week to Japan, where the sting is removed and it is turned into a kind of wafer.

In the meantime, the consumers of Spain and, indeed, the consumers of other EU countries, the United States and Japan have shown themselves ready to pay more, for reasons of diet-awareness and fashion, to eat fish. A sure sign of diminishing supply is that the price keeps going up. Fish has become more expensive in real terms as demand has increased. The rise in the price of seafood over the past 30 years is the more remarkable when you consider that the price of chicken, beef, pork and dairy products has fallen dramatically as a result of technological advances in farming.

For a decade or more, one thing had puzzled me, since I personally became convinced that the bounty of the seas must be shrinking everywhere. That was when Len Stainton, a fish merchant with a conscience, invited me to his office in Peterhead, Scotland's biggest North Sea fishing port, where he told me quietly: 'They used to land fish as big as a man. Now you are lucky if they are as big as your hand.' He took me upstairs to the boardroom and showed me pictures of the market in the early 1980s, when there were twice the number of boxes taking up twice the amount of floor space as there were that day. Fishermen, he told me bitterly, used to push fish off the pier if it failed to make what they considered to be the right price. This decline in catches, he said, must be going on all over the world. The thing that puzzled many of us at the time was that the decline happening before our eyes was not reflected in the annual figures for global catches produced by the ultimate authority on fishing statistics, the United Nations' Food and Agriculture Organisation (FAO). According to them, global fish catches kept going up and up. Perhaps there was more fish out there, somewhere, than people like Mr Stainton and I thought.

What we now know – thanks to a piece of superb detective work by Reg Watson and Daniel Pauly at the University of British Columbia – was that global catches, which had risen ever since 1950, had just then begun to decline. It took another 11 years for this information to become public, even though it was of vital importance to how we manage three-quarters of the globe and vastly important to the world's food supply. Perhaps most telling was that it appeared in the pages of *Nature* rather than being published by the FAO.

This is how the official figures were found to be wrong. The FAO had reported global fish catches increasing nearly every year from 40 million tons in 1950 to more than 80 million tons in the early 1990s. The upward trend continued, despite the collapse of Grand Banks cod in 1992. And despite the depletion of other fisheries, 75 per cent of which the FAO duly warned were fully or over-exploited, the total rose inexorably upwards to 95 million tons in the year 2000. So how come? Watson and Pauly looked at the productivity of the sea in every area of the world. The way they estimated it, productivity was consistent with the reported catches in every ocean but one. The waters around the world's most populous nation, the People's Republic of China, were known to be as overfished as anywhere else, but China went on reporting rising catches and a completely implausible total of 10 million tons a year. This, Watson and Pauly estimated, was at least double what was biologically possible. The reason for the misreporting was that officials in still-communist China get promoted only if they have increased production. So production miraculously increased. In fact, when more realistic estimates of Chinese catches were used, it became clear that global catches have been in decline since 1988. Watson and Pauly believe the decline is about 0.7 million tons a year, which is consistent with what we know about the drop in fish stocks around the world. An official figure for the decline does not yet exist, since the FAO, which now draws attention to the belief that China's figures are inaccurate, is still trying to persuade the Chinese government to provide accurate records of landings.

The discovery that the world is running out of fish, as its human population continues inexorably to increase, means that the human race has run into one of the limits to growth that environmentalists predicted it would hit in the 1970s. There are possible ways in which our food supply can be stabilised and increased through fish farming – though, as we shall see, predatory fish such as salmon depend on wild caught fish for food. What this means for the sea's wild fish populations, however, is that we have crossed a threshold from an era of complacency to one of concern. As Watson and Pauly put it in *Nature*:

> There seems little need for public concern, or intervention by international agencies, if the world's fisheries are keeping pace with people's needs. If, however, as the adjusted figures demonstrate, the catches of world fisheries are in general decline, then there is a clear need to act . . . The present trends of overfishing, widescale disruption of coastal habitats and the rapid expansion of non-sustainable aquaculture enterprises . . . threaten the world's food security.

I was talking to Donal Manahan, an Irish-born professor of biological sciences at the University of Southern California, who was an undergraduate at Trinity Dublin in the 1970s. He told me: 'I remember this lecturer in my first year saying, "In your lifetime all the world's fisheries will be gone. They will have collapsed or be in decline." I had no idea how right he was.'

CHAPTER 5
ROBBING THE POOR TO FEED THE RICH

DAKAR, SENEGAL, West Africa. The *Vidal Bocanegra Cuarto*, a 149-ton freezer trawler registered in Huelva, Spain, rode high on the black, oil-polluted water of Dakar harbour like a swan. The Spanish vessel had been painted from bow to stern in a light grey, probably on its most recent voyage, as tins were still lying on the deck. There was not a scrap of rust anywhere, except on the steel of its trawl doors and hawsers, which were constantly exposed to the sea. It stood out from the working Senegalese trawlers, with rivulets of rust across their paintwork, and from the many wrecks hitched together like pontoons for repair or scrap. Most corroded of these, and strangely sinister in the evening light, was the *North Sea I*, a listing, distant-water trawler dating from the era of the last cod war between Britain and Iceland in the mid-1970s. It was apt that this old buccaneer should end its days here.

As the press delegation landed to examine the Spanish trawler more closely, its skipper appeared and invited us aboard. Pepe José Vidal Acuna was rust brown himself from years at sea. He had put in to harbour for supplies and to land his catch – prawns, crab and lobster to be air-freighted that day to Spain's discerning fish markets. Mediterranean countries pay the best prices for prawns, a taste their home waters formed but, ironically, can no longer supply. Happy to be off duty, Pepe invited us to explore his vessel.

Bo Hansson, a Swedish fisherman and journalist, pointed admiringly at the array of screens on the bridge. The craft boasted two of everything: two fish-finders, two satellite navigation systems, two sonars, two computers, all to avoid a premature return to home port. The *Vidal Bocanegra Cuarto* lacked for nothing – she was a well-run, immaculate, fish-killing machine.

'The Spanish fish all over the world,' Pepe declared with an expansive gesture. Angola, Mauritania, Morocco. In Senegal his vessel fished with a 17-strong Senegalese crew and three Spaniards, who were clearly in charge. Pepe spent six months of each year catching shellfish off the coast of West Africa, and he caught enough on this long expedition to enable him to spend the next six months at home.

The day before in these same waters our motor-boat had pulled alongside a large *pirogue*, a traditional Senegalese sea-going canoe. The crew of 11 fishermen chanted songs as they hauled in their purse-seine net. It came in empty, except for a tiny broken sea horse and a few urchins. One of the young crew, Lamin Saar, showed us a scattering of sardinella in the hold: 'Today is very bad. We won't get anything except the price of the fuel.' He said that when his grandfather was fishing, *every* day was a good day. On a fortunate day now, each crew member can earn as much as £10, but this happens no more than once a month. The next day at M'bour, a large fishing port down the coast, a blue-robed elder in gold-rimmed dark glasses had come to meet the foreign press. He wanted to tell us his opinion of the deals signed by his government, which allow trawlers from the EU, Japan and Taiwan to fish in Senegal's waters. 'Poverty came to Senegal with these fishing agreements,' he said succinctly. Scientists show that there is more to this view than nostalgia.

Fed by one of the Atlantic's great up-wellings, the waters of West Africa are among the world's richest, with more than 1,200 species of fish. Upward currents, created by the force of trade winds rushing off the desert of Mauritania and out to sea, draw nutrients to the surface from the deep. The nutrients stimulate plankton, the base of the whole marine food chain. Senegal's fish markets, usually to be found on a shore strewn with brightly painted *pirogues*, provide evidence of this extraordinary ecosystem. Barracudas, 1.5 metres (5 ft) long and prized for their firm flesh, lie piled on tables. There are also pink African sole, sea bream, *thiof* or white grouper and *capitaine*, called in English giant African threadfin. There are strange, slab-faced fish and jelly-fleshed eels, which have no European name. Mothers hold up juvenile, plate-sized groupers, so beloved of the restaurant trade, that should not have been caught as they will not have spawned. Smiling children offer a few fish on a plate or a single octopus. There is something for everyone. The poor in Africa eat small fish, sardinella and other shallow-swimming, pelagic fish. These are caught by fishermen in *pirogues* with hand-held purse-seine nets and dried on raised tables by the women before being sold inland.

For about a decade scientists have been warning that the fish stocks of West Africa's continental shelf are over-exploited and that some, such as grouper and sea bream, are actively facing collapse. There is absolutely nothing in place to prevent their fate resembling that of the once-innumerable northern cod off Newfoundland or the North Sea cod. Hake, too, caught along with deep-sea prawns further down on the slopes of the continental shelf, is also overfished. If stocks collapse and local people starve, Europe, including the former Soviet Union which fished heavily in these waters until the end of the 1980s, will bear the largest share of the blame.

The European Union has had little success looking after fish in its own waters, but it has strong fishing traditions, and fishermen vote and know how to lobby. So the EU spends £127 million a year buying access for European fishermen to distant waters all the way from the Arctic Circle to the Falkland Islands. It recently signed a new deal with several West African countries, including Senegal. The beneficiaries of these dubious, little-publicised agreements, are distant-water trawlers from Spain, France, Italy and Greece. Among the countries the EU has signed deals with is Angola, where millions of people risk starvation. This is a matter of indifference to the country's elite, who earn hundreds of millions of pounds a year in oil revenues. They also receive £18 million from the EU for allowing 85 of its vessels to fish for tuna, prawns and demersal (bottom-dwelling) fish, which, from the EU's point of view, sounds like an extraordinary bargain. It is unclear whether British or German tourists on their package holidays in Spain, or indeed the Spanish themselves, are aware that the prawns in their paella are taken not from the Mediterranean, but from seas belonging theoretically to some of the starving in Africa.

Fishing in the waters of countries that are distracted by social unrest is a favourite trick of the EU fleet. The Spanish and French tuna fleets regularly fish in the waters of Somalia, the only country on Earth that doesn't currently have a government. The European Commission recently renewed its fisheries agreement with the Ivory Coast, which is in the throes of civil war.

There was a time when Britain also maintained a large distant-water fleet, sailing out of Hull and Grimsby, towns that vied for the title of greatest fishing port in the world. Then British trawlers fished for cod all the way to the Arctic Circle, in what became the territorial waters of Iceland, Norway and Russia. Now all but a tiny fragment of that industry is gone. The free-for-all ended with the signing of the UN Convention on the Law of the Sea (UNCLOS), which gave a legal basis to attempts by Norway and Iceland to close their waters to foreign fleets to conserve stocks – which in Iceland's case sparked three 'cod wars' with Britain. The Nordic peoples won. Iceland and Norway were too rich, too democratic to give away their birthright cheaply. African countries – or rather their ruling elites – have seen their waters more as a ready source of instant foreign exchange. So the neo-colonial days live on for Spain, which maintains a fleet of over 200 trawlers off the coast of West Africa, largely at other EU nations' expense.

The European Commission argues that fair access agreements are encouraged under UNCLOS, which established all countries' exclusive right to resources within their 320-km (200-mile) limits in 1979. The theory is that fisheries access agreements established a means for poor nations to profit from the harvesting of a surplus they did not have the technical means to harvest themselves. Such a surplus may theoretically still exist in the waters off Mauritania, a country with a relatively small artisanal fleet equipped with traditional vessels and a vast continental shelf, but not in the waters off Senegal. Senegal has an industrial fleet of its own, catching 48,000 tons of fish and shellfish a year, and a vast artisanal fleet, based on *pirogues,* which catches more than 320,000 tons. Fishing is a source of employment for 600,000 people, a significant proportion of the coastal population. This population is increasing as a decline in rainfall since the 1970s has brought more people from the interior of the country. As Calixte Ndiaye, a college professor who runs a small, local organisation that works with fishermen, told us, 'The sons of fishermen were taught to swim by the time they were two. Now you see many fishermen who can't swim.'

There is no longer a surplus of fish here. Daniel Pauly, professor at the University of British Columbia, who has spent years studying tropical fisheries, says that fish stocks off the west of Africa have declined by 50 per cent since 1945, when industrial exploitation effectively began. There is increasing conflict between industrial and traditional fishermen. The access agreements specify that only artisanal fishermen may fish within 10 km (6 miles) of the coast. But these rules are routinely broken. Industrial trawlers, both foreign and Senegalese, are known to venture within 10 km (6 miles) of the shore at night. The *pirogue* fishermen, equipped with modern nets and outboards, like to venture further out to sea. They concede that generally the Europeans tend to give them a wide berth if they are seen, but the Taiwanese often ignore torches lit by the sea-going *pirogues.* There are many fatal accidents – in 1997 nine fishermen were killed in a collision between an EU trawler and a *pirogue,* bought for poor villagers by a French development charity.

The argument of the EU and their partners to the Senegalese government is that the industrial trawlers and traditional fishermen are catching *different* stocks. It's not true. Foreign trawlers catch bottom-dwelling fish on the narrow continental shelf in up to 200 metres (660 ft) of water, the very same fish that traditional long-liners in their *pirogues* catch more selectively. It is true that foreign vessels in Senegalese waters do not catch the shallow-swimming fish, such as sardinella, that a fleet of other *pirogues* catches. These fish, however, migrate northwards to the waters off Mauritania, where they are caught by a fleet of vast, international factory ships. Monitoring in the Senegalese waters is rudimentary – there is one (frequently grounded) fisheries inspection plane – and there is plenty of scope to cheat the system.

The most telling point about the EU agreement with Senegal is that it imposes no catch quotas to conserve stocks. Instead it sets the total tonnage of vessels that may fish in Senegalese waters at any one time. The 150–250-ton trawlers in Dakar harbour can catch all they want provided they use the right mesh – smaller and less selective than in equivalent fisheries in the EU. The European fleet declares catches of 12,000 tons a year, but this is widely disbelieved. Reports for the WWF, formerly the World Wildlife Fund, estimate that the EU's catches had to be nearer 80,000–100,000 tons a year, up to eight times the declared total.

That is just the weight of the fish *landed*, not the ones that go over the side. Fishing for prawns also nets a number of fish (known in the trade as 'by-catch'). Fishermen who trawl prawns for Spanish paellas admit that they account for around 15 per cent of their catch. The other 85 per cent is fish, some saleable locally, much not saleable at all. The *Vidal Bocanegra Cuarto* fishes with a 40-mm (1½-inch) mesh net, and might expect to catch 20 kg (44 lb) of prawns – big creatures up to 20 cm (8 inches) long – and 50 kg (110 lb) of saleable fish per hour. The boat might expect to catch the same amount of undersized prawns, juvenile fish and inedible species, such as urchins, undersized octopus and crustacea, which would be dumped over the side. Rather disturbingly known as 'trash fish' by the industry, they include juveniles of all the most locally important fish species, particularly bream and grouper, the stocks in the worst trouble. So industrial trawlers like this can, justifiably, be accused of stealing Senegal's future. The prawns they fish for are declining, as well as the trash fish they catch in the same nets.

Jacques Marec, the French-born head of a fishing fleet of 19 trawlers based in Dakar under the Senegalese flag, observes that catches of prawns are declining overall by 300 tons a year; in 1983 each trawler was catching 150 tons of prawns – now it was catching 40 tons. This occurred partly because of increased competition, but mostly because of the absence of fish. Senegal's industrial fleet clearly shares as much blame for overfishing as anyone else, but Europeans cannot offload their responsibility that easily. The home fleet fishes almost exclusively for export to the European market. There's no escaping the fact that destruction of West African fish stocks has arisen mainly from demand in Europe, followed closely by Japan and Taiwan – a fact of which the consumers in those countries are almost entirely unaware.

The European Commission likes to proclaim that access agreements can be a way of helping the African countries involved. But it continues to negotiate them furtively, like a guilty secret. Its latest 'partnership' agreement with Senegal was conveniently negotiated just before the World Summit on Sustainable Development in Johannesburg in 2002, where the EU undertook to implement recovery plans for endangered stocks by 2015. The Senegalese deal was brokered, with scant publicity, six months before ministers agreed to improve the terms of such agreements in future. Dr Ndiaga Gueye, director of marine fisheries for the Senegalese government, has no illusions about what he was negotiating. 'We are talking about commercial agreements,' he said. He confirmed that in the course of an 18-month negotiation, which he headed, the EU actively resisted numerous conservation measures and drove a hard bargain on price. He declared that under the terms of the agreement, Senegal would earn far less if it were to impose quotas to save its endangered stocks. The Japanese had provided Senegal's single research vessel, while the EU had provided no help whatsoever in managing its stocks.

So why did Senegal sign? Dr Gueye indicated that the money, £42 million, was a tidy sum to a poor country. Senegal needed hospitals and schools so that the next generation might learn to be something other than fishermen. He claimed that Senegal was open and transparent about how these revenues were spent, but it still didn't seem to add up. So I asked what else was in it for them. Dr Gueye spoke obscurely of 'political and diplomatic reasons'. Did EU negotiators make a link between fisheries and aid? That is the clear impression he gave. We also learnt that diplomats from the Spanish embassy in Dakar – interestingly not Brussels officials – spent a lot of time at the fisheries department before the deal was finally agreed.

The local fishermen do like some things in the new agreement. A proportion of all crews must be Senegalese and, instead of being shipped straight out as before, 17,000 tons of tuna a year – caught off Senegal, The Gambia and Mauritania – must be landed in Senegal for processing. Tuna is the one species that the local fleet does not have the means or the experience to catch. All the same, Senegalese trawlermen question the legality of the access agreement now that stocks are clearly overfished, and the artisanal fishermen are angry that the agreement was signed behind their backs. Environmentalists say that 'cash for access' deals signed by the EU still stink but are better than previous ones. Julian Scola of WWF says they raise 'serious questions about whether these agreements are compatible with the EU's rhetoric on sustainable development and with its development policies which are about eradicating poverty'. Environmentalists are caught. They cannot oppose access agreements altogether because if the EU were to pull out, it would open the door to murky deals with individual companies or other fishing nations who might behave no better. Instead, they want the terms rewritten to promote conservation.

What makes conservation even more difficult is the growing number of EU trawlers transferring to the Senegalese flag. Local boats face less stringent rules, carrying no official Senegalese observers, as larger EU vessels are obliged to do. In theory, joint ventures between local companies and foreign fleets have to be 51 per cent Senegalese-owned. But no one really complies with this rule. 'There are people who one day have problems buying long trousers and the next day are the owners of four trawlers,' one Senegalese official remarked tartly. All the observers I met had been offered bribes by EU skippers to turn a blind eye to the rules. The

EU has said it wants the transfer of EU vessels to foreign flags to stop, but characteristically it has given them a year's grace and, when introduced, the new rules will not be retrospective. Needless to say, joint ventures are proliferating rapidly.

'People have got to recognise the new situation,' says Jacques Marec, president of a Senegalese trawler firm. In this industry a new wind of change is sweeping through Africa, and two fisheries access agreements – at last recognised as the neo-colonial practices they were – have been annulled. Namibia was the first to crack down on EU trawlers in 1992. It was followed in 2001 by Morocco throwing them out altogether. Spain was furious. Even after Namibia became independent in the 1970s, Spanish trawlers continued to harvest huge quantities of the country's hake stocks. By then the hake was almost all classed as juvenile (below spawning size), and there was an acute danger of stock collapse. The Namibians hired a helicopter, and local fishery inspectors alighted on the decks of the illegal fishermen to carry out arrests. The Spanish were indignant, but gradually the Namibians took control and used their marine resources for their own benefit. Fisheries have since become that country's main engine of economic growth.

Ever resourceful, though, the EU persists in signing up new fisheries agreements with countries that don't know or don't care what they are letting themselves in for. It recently signed its first agreement in the Pacific with the Solomon Islands. Somebody in the Spanish or French distant-water tuna fleets, which have the only vessels capable of dashing to the western Pacific on whim, must have been reading the latest assessment of the world's fish stocks by the UN Food and Agriculture Organisation. This says that the western Pacific is one of the few places on Earth where fish stocks are under-exploited. We'll just have to see how long that lasts.

Senegal, in many ways a model democracy by African standards, has some unenviable decisions ahead. It recognises that its stocks are endangered and that something has to be done, but this has yet to be matched with urgency or action. The consensus is that Senegal has five years to restrain the growth of its artisanal fishing or face disaster. Even if it ejects foreign trawlers from its waters, it faces challenges restraining the growth of its own industrial fleet, a growing proportion of which formerly flew European flags. The future doesn't look good: globally there are few successful models for managing such a complicated mixed fishery. There are many other African nations in the same position as Senegal – starting with its neighbours in West Africa, but including countries in the Indian Ocean – and countries all over the world who rely on their 320-km (200-mile) limits as a major source of income. Through model agreements and more informed buying habits, European voters and consumers could exercise important pressure for the conservation of fish stocks, but the European Commission listens only to vested interests and talks of good governance while bribing Africans to persist with unsustainable practices and to allow the pillaging of their waters by EU vessels. The present access agreement with Europe has only hastened the onset of disaster.

Here ends this book's headlong plunge into the disturbing things going on in the world's oceans. You might find that your head is spinning, that there is too much to take in, more than one can possibly feel worried about as a consumer or voter. I confess that there are times when I do as well. When this happens I try to reflect upon a single sea – the muddy, industrial North Sea that begins 5 km (3 miles) from my door. It is one of the most heavily used seas in the world. It supports a major oil and gas industry; it was once both the fish basket and main highway for northern European civilisation; it is also, though we often forget this, an ecosystem full of wild creatures, and one that has changed considerably. When I begin to reflect on what has been lost from the North Sea I become resolute. I feel the anger rising again.

CHAPTER 4

SEA OF TROUBLES

The produce of the sea around our coasts bears a far higher proportion to that of the land than is generally imagined. The most frequented fishing grounds are much more prolific of food than the same extent of the richest land. Once in a year, an acre of good land, carefully tilled, produces a ton of corn, or two or three hundredweight of meat or cheese. The same area at the bottom of the sea in the best fishing grounds yields a greater weight of food to the persevering fisherman every week of the year.

Professor Thomas Henry Huxley
Address to the International Fisheries Exhibition, London,
1883

TEN THOUSAND YEARS ago, one of the most productive seas ever for fish – the North Sea – was tundra. The last ice age suspended so much water above ground as glaciers and snow that sea levels around the world were about 50 metres (165 ft) lower than they are today. What is now the southern North Sea was part of continental Europe and settled by humans. Tribes from the Low Countries used to cross what is now the sea and follow a funeral route along the Ridgeway to sacred places in southwest England. Sites of Stone Age coastal settlements have been found on what is now the sea bed, their middens (rubbish heaps) containing shells and the bones of fish and sea mammals. The archaeological record shows that early Europeans clung to the shore because the sea was their principal source of food. A diet rich in marine mammal oil would have helped to keep them warm.

The North Sea is therefore a shallow sea, two-thirds of it under 50 metres (165 ft) deep – about the same as the height of Nelson's Column in Trafalgar Square. Think of the plaice, if you will, as pigeons. One of the first things any visitor notices about the southern North Sea is that it is muddy. The waves are brown-grey in an easterly gale, and blue-green with suspended sediments and plankton in high summer. Visibility for divers is seldom more than a few metres. This is because of plankton growth and because sediments carried down by the rivers and the sand and gravel of the bottom are churned up by the waves and tides. Locally the water can be even muddier as a result of dredging to keep ports clear and to extract sand and gravel. The water clears briefly for about three weeks in May, but in the southern part of the sea, that's about it.

The northern part of the North Sea is deeper, dropping to more than 700 metres (2,300 ft) in the Norwegian Trench. In the northern part of the sea, divers will tell you, the water is clearer and the bottom harder – there is what scientists call a hard substrate, which supports a greater range of plants and plant-like animals, together with a greater variety of shellfish. But was the southern North Sea always muddy? In all the time I have spent holidaying on its shingle or sandy shores, swimming in it, sailing across it or watching ferries, rig tenders and fishing boats in the evening light, it never occurred to me that there might have been a time when the sea was not cloudy. Only recently have I discovered that the muddiness of the water may well be explained by overfishing.

Vast natural oyster beds covered large parts of the east coast of England at the time of Christ, and were traded throughout Europe even before the Romans invaded Britain. Remnants of these oyster beds remain productive in Essex and Kent, but at a much reduced rate owing to centuries of fishing and a parasite that devastated the population in the 20th century. What is not often remembered is that there were also vast oyster beds in shallow areas of the open sea, producing 100 times more oysters than today only a century ago. Nineteenth-century maps show oyster beds 200 km (120 miles) in length on the German and Dutch side, but the last of these were fished out before the Second World War. Since then there have been no oysters left to form a hard substrate across the bottom. Maybe there have been changes in the Atlantic water entering the North Sea that have made the sea bed a less comfortable habitat for oysters. But the simplest and most obvious explanation for the disappearance of the giant North Sea oyster beds is overfishing.

With a covering of vastly more oysters, many of them very large, and a bed of the shells of former oyster generations, the bottom of the North Sea would have been less mobile and its plankton more in demand as food. The bottom would have settled into a hard substrate and the plankton and suspended sediments in the water would have been stabilised by the bivalves as food and glutinous excrement. The hard bottom would have supported lobsters, today less plentiful in those parts of the sea. It therefore takes only a short step to conclude that the water was probably clearer, certainly when one adds to the filtering power of the oysters the larger mussel beds that would then have existed around the North Sea coasts, and the greater areas of salt marsh and shallow estuary that would have caught more of the sediments coming down the rivers. If we assume the southern North Sea was once clear, it follows that increased sunlight would have reached the sea bed, improving the productivity of the ecosystem generally, but in particular the bottom flora. It is probable that the vast areas of mostly featureless sand and gravel at the bottom of the North Sea represent a devastated ecosystem, its natural purifying capacity lost.

Environmentalists from North Sea countries are fond of complaining about eutrophication – the excessive growth of plants and algae, some of which are toxic, because of excessive nutrients dissolved in the water. While it's true that high levels of nitrogen and phosphate come from farm run-off and sewage plants, the influence of overfishing oysters is never mentioned – probably because the groups in question don't know it happened. They, like the rest of us, tend to assume that the sea they experienced in child-hood was 'natural' and as it should be. But studies of Chesapeake Bay on the other side of the Atlantic, now a eutrophic soup of plankton, show that it was once clear. Oysters the size of dinner plates used to eat enormous amounts of plankton, filtering the water of the bay once every three days. In the 1700s, the oyster reefs were so large that they were a hazard to navigation. Needless to say, these oysters were fished out. The sedimentary record is likely to show that parts of the North Sea had the same filtering capacity. Now, it is unlikely that oysters would re-establish themselves, even if they tried. Oysters need to be undisturbed for four years to reproduce, and most suitable parts of the southern North Sea are trawled at least once a year.

Also missing from the sea, though present historically, are a number of large mammals. Largest of all was the grey whale, which feeds on mussels on the sea bed and in estuaries. Mussels and oysters would have lived in huge profusion in the delta of the Rhine and in the Wash, especially before land there was 'reclaimed' from the sea. Grey whale bones have been discovered around the North Sea, the most recent dating from the early 17th century. There were right whales – actually baleen whales that swam slowly and floated when killed, so they were the 'right' ones to catch – at least until the late Middle Ages. It is fair to assume that there were more dolphins, common and bottlenose, and harbour porpoises, but fewer seals than there are today. The laboratory director of the Royal Netherlands Institute for Sea Research on Texel Island off the northwest coast of Holland recorded seeing six harbour porpoises at the same time from his office window in the 1950s. Now there are hardly any harbour porpoises in the southern North Sea. I once became dreadfully seasick while on an acoustic survey intended to find them, which concluded that there were likely to be none south of Flamborough Head on the Yorkshire coast. Records of sightings from ferries to and from Texel Island document bottlenose dolphins that are not there today.

So much has changed, even in a single lifetime. It becomes an awesome task to document what has happened over millennia, both naturally and as a result of human intervention. Traces of even more distant times are present in the North Sea. Trawlers still pull up bones of mammals from the Pleistocene era – bison, musk ox, woolly mammoth and woolly rhino – and a trade exists in sorting these for museums and private collections. Within our own climatic era (the past 10,000 years), 'haul-ups' of walrus by nomadic people would certainly have happened, and may be assumed to be the reason we no longer have that kind of sea mammal. We know porpoises were hunted by Stone Age people who lived and beached their boats on a shoreline that is now several kilometres off the Danish archipelago. Sailing boats, harpoons, traps and simple trawls may on their own have eradicated some slow-moving species of fish and mammals, just as other oceans lost their turtles and sea cows.

Inevitably, it is easier to list what has gone in the past two centuries. When it comes to fish, most of the absentees are those that migrated from salt water to fresh water to spawn, thus multiplying the risks to their survival. The sturgeon is one such migrant. Sturgeon are living fossils, much more common 60 million years ago than they are now. Atlantic sturgeons were caught in the Netherlands until the 1870s, when inventive fishermen employed steam engines to drag nets along the river bed. The sturgeon of the Rhine delta are now gone. The houting, a migratory, cod-like fish, is another absentee from the western side of the North Sea, found occasionally on the Continent. The Atlantic salmon, now in decline, must once have been as plentiful in rivers on the east and west coasts of the North Sea as it is on the Kola peninsula in Russia today, roughly as plentiful as the Pacific salmon species are in Alaska. There it can still be said that you could almost walk across the water on their backs. At one time that must have been true of the Thames and Rhine.

Perhaps the biggest fish absentee, the bluefin tuna, disappeared as recently as the 1950s. Tunny fish, as they were called, used to follow the herring shoals. Specimens of stupendous size were caught on rod and line within 40 km (25 miles) of Scarborough from 1929 until 1954. Then, for reasons that have never been satisfactorily explained, the run disappeared. The pioneer of the rod-and-line sport was one L. Mitchell-Henry, who dedicated himself to catching a giant bluefin after a large fish was harpooned off Scarborough in 1929. He eventually landed the British record, a fish of 383 kg (851 lb), from a rowing boat in 1933. The sport was such a draw for Scarborough's tourist trade that the council made a club-house freely available to the British Tunny Club, formed in that year. Mitchell-Henry fell out with the club when the sport was overwhelmed by rich socialites, who used large boats and what he considered to be less-than-sporting techniques. The end of a privileged era came with the Second World War, but not before Captain C. H. Frisby, VC, had set a world record for the greatest weight caught in a day, with five tunny weighing a total of 1.25 tons in 1938.

British people had yet to acquire a taste for raw fish, or even tuna steaks, in the 1930s, so most tunny were sold for fish-meal. Wasteful as this sounds, the sport fishery for North Sea bluefin is unlikely to have caused the tunny's demise. British government scientists say that it was caused by climate change or industrial fishing of the bluefin off the African coast. This explanation remains unconvincing, as there is still a small migration of large bluefins off the west coast of Ireland and into Norwegian waters. Indeed, a bluefin of 562 kg (1,241 lb), 241 lbs larger than the 1,000 lb tunny dreamed of but never landed by the big game anglers of the 1930s, was caught in a trawl off the Irish coast in 2004. How much longer the run off Ireland is likely to survive is debatable now that a sport fishery has been established exporting its catch to Japan.

More certain, and extensively recorded in the past century or so, has been the effect of fishing on what are called commercial stocks – a description that avoids considering cod, haddock, plaice and sole as wild animals, which is what they are. The greatest impact of commercial fishing is on long-lived species that reproduce slowly, whether or not these are the fishermen's prime targets. No doubt there were once more species of shark and dogfish in the North Sea, but these are likely to have been caught as by-catch. Within the last 100 years there were common skate in British waters, which, when hung by the gills, were the height of a man. Barndoor skate, an even larger species which lives on the eastern seaboard of North America, have died out too. In the southern North Sea, the common skate has declined by at least 99 per cent since pre-industrial times, and is probably extinct in the North Sea, but it can take 50 years or more to prove that. It is provisionally listed as 'endangered' by the International Union for the Conservation of Nature. Public awareness of a potential extinction, for which the public is partly to blame, is not high. Restaurants and fish and chip shops have long found a substitute in thornback ray, which is helpfully labelled as skate because that is what the public expects.

As to the main fish species, the principal change, of course, has been in relative abundance over the past 50 years, which is likely to have all sorts of knock-on effects within the food web. While preparing to write this book, I had a battle on the telephone with a scientist from ICES over the question of what the virgin stock of cod in the North Sea was, or even what it was before big sailing trawlers geared up to fish for them in the 18th century. You would think that after a century of studying the sea at public expense, and at a point where we badly need such knowledge to make decisions for the future, the scientists of ICES would have some idea, but no. Learned committees are looking into the matter and may take five years to pronounce. ICES refuses even to hazard a guess at how many cod there once were, which is convenient since a high figure would throw into greater relief the disastrous decline in cod numbers that ICES has overseen. Ransom Myers, of Dalhousie University, Canada, has estimated that North Sea cod stock was once 7 million tons, though some feel this is likely to ignore depredations by other species. One model said the cod, even now, would probably recover to between 400,000 and 600,000 tons if there were no fishing.

We do know that spawning stock of North Sea cod in 2003 was estimated at 53,000 tons, so we are entitled to conclude that we have lost around 90 per cent of the North Sea cod that 'should' be there. Reduction to less than 10 per cent of the original spawning biomass happens to be how the Canadians define population collapse. When this happened to its cod stocks, Canada banned cod fishing. The EU, on the other hand, has spent four years talking about it, and we are still catching cod. This, even the reserved, pro-fishery scientists of ICES realise, is far from sensible. But the North Sea is what is known as a 'mixed fishery', with a far greater variety of commercial fish species than the Grand Banks. In other words you cannot stop catching one species without curtailing fishing for something else. In 2003, for the second year in a row, ICES called for a ban not only on fishing for cod, but on fisheries that catch cod in their nets as by-catch. It did this without any great expectation that North Sea ministers would be able to steel themselves to agree to such a thing.

When the age of the steam trawler was at its height, the Victorian scientist Thomas Huxley famously declared that an acre of good North Sea fishing ground produced a ton of fish a week and an acre of land a ton of grain a year. Huxley meant sizeable cod, haddock, whiting, plaice and sole landed into Hull, Grimsby or Lowestoft. The productivity of the sea in those terms is now about a tenth of what it was in 1883 because the populations of the other fish have declined as much as the cod. Huxley fell victim to a number of misconceptions about the productivity of the sea. An acre of the best ground, of course, happened to be where fish gathered: it was the whole of the North Sea ecosystem, a much larger area, that was actually feeding them. His comparison with farming overlooks a crucial difference between farming and hunter-gathering: fisher-men reap but they do not sow. Investment in new technology has the opposite effect in farming to that in fishing. An acre of Lincolnshire clay, bought in many cases by successful fishing families, would each year produce a ton of wheat to the acre in Huxley's day. It now produces 5 tons of wheat to the acre because of advances in technology. The trajectory of marine hunter-gathering has been in the opposite direction because the extraction rate was probably unsustainable at the time and has been made worse by every technological development in fishing.

The autumn-spawning North Sea herring lays its eggs on gravelly areas of the sea bed, but its spawning grounds are actually located in the beds of ancient rivers, and some groups of herring still spawn in inshore waters and estuaries. This suggests that the herring once evolved to live in rivers but later adapted to the marine environment. The plentiful herring has been a source of conflict and war for centuries. The traditional drift-net fishery which moved from Shetland to the English Channel with the herring shoals through the season employed huge numbers of sailing craft in the 19th century. The herring enjoyed a recovery during the Second World War, but its spawning stock fell from 5 million tons in 1947 to 1.4 million tons by 1957 as the traditional drift-netting was replaced by trawling on vulnerable spawning aggregations in the Channel and around the Dogger Bank, where the sheer intensity of fishing is thought to have disturbed the spawn on the gravel. By 1975, when the spawning stock had fallen to 83,500 tons, the herrings' spawning grounds around the Dogger Bank were no longer used. They are disused today and the autumn drift-net fishery on the East Anglian coast is gone, too. The herring, however, is a rare example of the success that comes when politicians actually take scientific advice. The collapse of the stock in the mid-1970s led to a ban on fishing for four years. The stock recovered, but not to its post-war level. It started to decline again in the early 1990s, and drastic action was taken again in 1996 when quotas were cut by half. Since then the stock has been rising slowly and the trend is still upwards.

With that exception to the rule, the present situation for table fish is bleak. The North Sea mackerel collapsed in the 1970s and has never come back. The major bottom-living species – cod, plaice, monkfish and sole – in the North Sea are all now listed by ICES as 'outside safe biological limits', meaning that their populations are so low that they could suffer breeding failure on a large scale. The sea is almost empty of older fish. A plaice gets no older than six years nowadays, but, given the chance, can reach the age of 40. A cod, likewise, will rarely reach six years old, but has evolved to live 20 years and more, laying larger, healthier and more numerous eggs when it does so. Fish have evolved to deal with the effect of climatic variability on their reproductive success by simply living longer, until more favourable conditions return. With fishing pressure as high as it is today, and few if any refuges in the sea, this option is closed off. Fishing pressure has even caused the cod and haddock to breed a year earlier than they did when fish survived for longer in the sea, a rare example of human-induced evolution. Worryingly, there are also other signals coming from the sea. Cod, haddock, whiting, sole and plaice are growing less quickly than they did in former decades. The cold water copepod, a form of zooplankton that forms a part of the cod's diet, has moved 600 miles northwards up the Bay of Biscay over the past decade. The question now is whether warming sea conditions – which have already led to the spread of southern species, such as red mullet and anchovy, into the North Sea – could remove fish, such as cod, from the food web altogether.

Huxley was almost right about one thing, namely, the phenomenal productivity of a shallow sea. The wanton destruction of palatable fish in their millions does not mean that there has been a decline in overall biological production. While the overall tonnage of palatable fish has reduced by a huge amount, the tonnage of life in the sea – what scientists call the biomass – is probably very much the same. What has changed is the relative balance between species. The sea has not been emptied, as some descriptions of overfishing misleadingly suggest. This understates the unprecedented nature of what is going on.

What has happened is that predatory fish have been removed from the system. We are not even sure whether the system will allow them back – whether what we are seeing is a one-way 'ecosystem flip'. The greatest weight of living creatures is no longer long-lived animals, such as the skate, cod or oysters, but small, short-lived creatures, such as prawns, langoustines, Dublin Bay prawns or scampi, sand eels, starfish, jellyfish, worms and plankton. Overfishing has emptied the sea of many of the animals that are most attractive or delicious to humans. A few thousand tons of what is left in the sea, such as prawns and langoustines, is of quite high value. A few hundred thousand tons has its use as feedstock for salmon farms, perhaps more if plankton can be turned into food. But overall we have changed life in the sea into something less 'natural', less useful, something potentially more hostile in times of crisis or disaster if we had to depend upon it, as, for example, Stone Age hunters did. This has been described as 'fishing down the food chain'. Some fisheries scientists treat the change caused by such fishing as neutral and say it is for society to say whether it is good or bad. I would say that most of society is ignorant of these things – partly the result of those scientists' failure to communicate the magnitude of what is happening to anyone except other scientists. When society has the opportunity to register its opinion about its environment it routinely concludes that it wants the environment to be as 'natural' as possible.

Trawls and dredges, of course, do not just affect the species they are designed to kill. They affect the whole flora and fauna of the sea bed. Without trawling, Huxley's acre of sea bed would support a community of plants, plant-like animals called hydroids, bryzoans, worm-tubes and a variety of shellfish life, many of them important prey for fish such as cod. In the western English Channel, which is clearer, the sea bed supports brilliant coloured corals and sea fans. Just north and west of the North Sea, in Scottish and Norwegian waters, there are reefs of the cold-water coral, *Lophelia pertusa*, the northern ones quite shallow, the southern ones at great depth. Some 40 per cent of these cold-water reefs surveyed in Norwegian waters were found to have been extensively damaged by trawls. It is calculated that these could take 100,000 years to grow back. Fishing with bottom gear also leads to the destruction of shellfish and seagrass beds, maerl (ancient calcified seaweed) grounds and fragile reefs built of tubular worm-casts.

Imagine a 2,000-horsepower trawler targeting plaice or sole, pulling a beam-trawl mounted with rotating tickler chains across even the more featureless sands and gravels of the southern North Sea. The weighted trawl and its chains, designed to beat flatfish out of the sheltering mud, smashes everything it does not catch, particularly the burrowing animals in the sediment, disturbing them at depths of up to 20 cm (8 inches). EU scientists have calculated that up to 7 kg (16 lb) of dead marine animals are killed by beam-trawls to produce 450 g (1 lb) of marketable sole. Sea urchins, hermit crabs, brittle stars and razor shells are killed or, having part of their shells broken, made vulnerable to predators. Starfish tended to be the survivors, losing a leg or two but escaping the net. The winners in heavily trawled areas tended to be crabs, which moved in from other areas to eat damaged creatures and worms that took over from the easily damaged bivalves.

A study of the bivalve *Arctica islandica,* which is thought to live for up to 150 years, found that it was easily damaged by trawling. Those that were badly damaged tended to end up in the stomachs of cod and other fish, thereby reducing the population considerably. Damaged ones, however, were found to be able to repair their shells, though grains of sand became bound up in the growing matrix. By studying annual growth rings and grains of sand in the shell matrix, scientists concluded that most parts of the southern North Sea were disturbed by a beam-trawl at least once a year. Some of the effects of trawling may be quite short-lived: for instance, the tracks of a heavy trawl, or scallop dredge, persist for up to two years. With some major sea-bed structures, such as corals, the damage is longer lasting. Many parts of the southern North Sea are thought to have once been covered with an extensive network of reefs up to 50 cm (20 inches) high built by the calcareous tube-building worm *Sabellaria spinulosa.* These are now extremely rare as trawling easily destroys the reefs. A surviving example is to be found in one of the British gas fields, an area where trawling is difficult because of the danger of nets becoming entangled.

So what has had the most extensive impact on the seas – fisheries or pollution? I lived for more than three decades believing it was the latter, in common with the environmental groups of the day, until I walked by mistake into a lecture at a North Sea conference in The Hague in 1990. A Dutch scientist called Han Lindeboom was presenting his findings, and he claimed that the effect of commercial fishing methods on fish is clearly more fatal than pollution because we record every year the millions of tons of fish these methods kill. In general, pollution effects are local, whereas fisheries cover the whole of the North Sea – and all continental shelf seas. Lindeboom recently updated some of his calculations. He now finds that the impact of fisheries on bottom-dwelling animals is a thousand times higher than that of sand and gravel extraction in the Dutch part of the North Sea. He also finds that the damage caused by fishing is 100,000 times higher than that of oil or gas exploration. The reason for these findings is that the extraction of aggregates and oil and the explorations for gas take up only small areas, whereas nowhere in the whole North Sea, other than directly under an oil rig or more recently a wind farm, is permanently closed to fishing.

CHAPTER 5

MIGHTY SEAMAN

CAMPBELTOWN, MULL of Kintyre, Scotland, 6am. Tommy Finn, skipper of the *Gleaner*, an 18-metre (60-ft) trawler based in one of the remoter parts of the Scottish mainland, steers us out into a force seven wind as it gets light. The shipping forecast says the weather will worsen, so we pass boats coming in. There are 'white horses' outside the bay, so they are not going to risk rough weather. The steel-hulled *Gleaner*, with a 400-horsepower engine, can withstand more of a buffeting, though Tommy informs me that the heads (toilets) are out of commission and provides a bucket in case nature calls. Whether it is the thought of this or the smell of the fry-up being prepared by Hamish, the cook to the four-man crew, I already feel green about the gills. It takes me the rest of the day to find my sea legs.

I am aboard the *Gleaner* because I am curious. I have never been out in a trawler off the west of Scotland, yet I have stared out at trawlers and inshore boats for years from the Isle of Arran, where my Scottish wife's family lived and where we spent Christmas and summer holidays. At those times, when I went for a walk up the hill, I noted that I could count six fishing boats from any high point, either facing out towards the Mull of Kintyre, or inland towards the mouth of the Clyde. How, I wondered, could there be enough fish for all these boats to catch?

As a child, I read about the sea angling festival off the Isle of

Arran, the oldest and most famous event of its kind in Scotland, which attracted competitors from all over the country. Prodigious catches of skate, tope, halibut and large cod were landed in the 1960s and early 1970s. The winning bag was a ton and a half. Over the past decade or so, the event fell into decline, the top catch amounting to only a few pounds, so it was given up. I mention this to Tommy and say it is an indication of the deterioration there has been in the Firth of Clyde, but he disagrees. As a professional who fishes all over the Irish Sea, he says there are still plenty of fish out there to be caught by those who know how. He says it is possible the amateur fishermen have just exhausted the wrecks and marks they know.

Tommy is the son of a leading fisherman, and is known as a particularly resourceful and enterprising fisherman himself. In a sea where stocks of cod have virtually disappeared and where fishermen who followed the herd have gone out of business, he has turned the business of catching fish into a science. When nobody realised that haddock had returned to the Irish Sea in quantity – the principal quarry there used to be cod – Tommy turned to catching them. When he realised there was a market in chip shops for dogfish, he caught them instead. Indeed, he was more successful than he was prepared for on one occasion. He caught so many dogfish in one haul – 44 tons – that he could not haul the net out of the water, so he had to tow it into port. That picture made the front of *Fishing News*. He says he didn't lose a single fish from that haul and sold the lot for £40,000.

Tommy is a skipper who believes in doing things differently. Instead of using an 'otter trawl', where steel 'doors' plane through the water like kites to keep the net open, he uses a method called Danish seining. This involves dropping off a buoy attached to 2½ km (1½ miles) of line, then shooting the net. The boat then turns around and steams back while paying out a similar length of line attached to the net until, finally, both ends are attached to the winch, the winch is engaged and the boat draws forward in low gear. The trawl ropes, laid out in the shape of an ellipse, dance inwards over the sea bed, herding fish into the trawl mouth.

Tommy explains that fish never swim outwards through the disturbance of silt created by the encroaching trawl lines. One wonders how he knows, other than because it works. Since the method covers a wider area than a simple trawl, it is arguably just as efficient at catching fish as a much larger, heavier trawl. It takes less horsepower to haul, which means a big saving on fuel. Danish seining therefore minimises a boat's costs and maximises the fishermen's share, which is divided equally after an allowance for Tommy, the owner of the boat.

The first net is shot just southeast of the Mull, and the haul begins as the sky brightens; brilliant foam bursts past us on the fish deck and the air begins to fill with wheeling, opportunistic gannets. The gannets, by nature solitary divers, roost nearby on Ailsa Craig, the volcanic plug known to travellers as Paddy's Milestone because it marks the way between Glasgow and Belfast. The gannets have learnt to home in when a vessel hauls its nets. The cod end – the toe of the net where the fish end up – comes in sight, with a respectable few stones of fish, mostly haddock, which are the intended catch. As the net comes in slowly, John and Paul disentangle the sea urchins, seaweed and isolated fish. The cod end is opened on the deck and a miscellany of creatures spills out. The haddock are sorted into boxes and will be gutted in spare moments between hauls, thus adding value because 'roundfish', as ungutted fish are known, attract lower prices. About a third of the catch is sluiced over the side. This includes fish that were the wrong size to be landed, known as 'discards', and creatures with no commercial value, known as 'by-catch'.

It seems astonishing what variety is in the haul, but nothing too exotic. There are dabs, a small turbot, a couple of monkfish, small 'hounds' or dogfish, juvenile cod and haddock under the landing size, then there are gurnards, a dragonet, a mullet and two small octopuses, both of which are clearly still alive and well, unlike most of the fish, which are flapping in distress with their swim-bladders ruptured, and unlikely to survive if they go back. Paul puts what he thinks Hamish will want to keep for the pot into a red box, a contrast from the white boxes in which the catch will be iced. The rest go overboard, into the mouths of hundreds of wheeling gannets. I watch but none of the discards or by-catch survives.

By-catch and discards are a fact of life to a fisherman. There is no fishing method that catches only the quarry. This kind of trawling is relatively 'clean', but still troubling to someone who is not acquainted with trawling. Some fisheries and times of year produce an even larger by-catch. Some prawn fisheries regularly have a by-catch of 85 per cent, though some of the palatable fish caught as by-catch will be sold. The UN Food and Agriculture Organisation estimates that about a third of what is caught worldwide, some 27 million tons, goes over the side. This takes what is hauled from the sea to around 120 million tons a year. Add to that the number of organisms that are killed or damaged by net, line or trap and are never landed – such as whales, porpoises, turtles and birds – and the number of animals destroyed on the bottom, and the total catch by fishermen reaches something more like 200 million tons a year. Consider that much of the weight of a palatable fish is head, cartilage, bone and offal, which goes over the side or is thrown away by processors. Consider also that about 40 million tons of fish is caught to make industrial products and food for farmed fish. Consider that some of the palatable fish caught will be turned into products for non-human consumption – as cat food, for instance. Consider that there may be an element of waste because some fish will not sell. Taking all these things into account, it is possible to conclude that the amount of protein eaten by someone or some- thing is maybe less than 20 per cent of the 95 million tons landed, and only 10 per cent of the amount of marine animals destroyed annually in the oceans. These are rough figures, but, given a wide margin of error, they are about right. So catching wild fish is a wasteful business. Today, aboard the *Gleaner*, you cannot but be impressed by the productivity of the sea.

Next we are beset by frustrations. The net comes in torn after the second haul. Although we saw nothing on the sounder, and Tommy's plotter has guided us to the exact place where he caught haddock a year ago, there is clearly some projection down there, a rock or an anchor. Tommy leaps down from the wheel-house, grabs a needle and directs the repairs being carried out to the net, leaving his second skipper, Lachlan, steaming to where he is to drop the next line. Then the radio picks up an emergency beacon from a fishing boat they know. It is fine weather now and the wind has not risen as expected, so there is no obvious reason why the boat might be in trouble. All the same, Tommy holds a conversation with the coastguard, then heads off on the bearing he is given at top speed. Halfway there, somebody manages to raise the boat sending the distress signal, and its skipper apologises, saying he was below working on the engine and the beacon had gone off by mistake. We return to the business of catching fish, but a net snags again, and this time we have to go back to pull it off.

When Hamish's generation began to fish after the Second World War, all they had to find the fishing grounds were compass coordinates on the back of a cigarette packet. Now, using satellite information from the Global Positioning System (GPS), Tommy plots three more trawls on lines that were successful at the same time last year. Since the end of the cold war, the GPS satellite, designed originally for military and naval use, has become even more accurate because the USA has improved the resolution of the signals. This means there are fewer no-go areas for boats, and they can fish within less than 10 metres (33 ft) of known rocks and obstructions. For most of the day I am out with Tommy, there is also a comforting grey cloud on the bottom on the sounder screen, an indication of fish.

Now the crew is rapidly gutting away on the fish deck, and a moment of decision is upon us. When you add today's catch to what has been on ice in the hold overnight here are nearly 80 boxes of cod and haddock on board. Tommy holds a consultation on his mobile phone, hears that prices are holding up in Fleetwood and decides to head for home rather than stay out another four days as originally intended. A white van is waiting as we tie up at the quay. We are all looking forward to an unexpected meal ashore, a drink and a night in a bed rather than a bunk.

That was seven years ago. Tommy is still fishing, mostly with the same crew. There is not much else to do in Campbeltown, except for working in the seasonal tourist trade or for a new company making wind turbines. The name *Gleaner* now belongs to a bigger boat – 24 metres (80 ft) – that Tommy acquired last year. He says there were fish in the sea and he thought it worth gearing up. He sold the old boat well, and the bigger one was cheap because so many people were trying to get out of fishing. He is pair-trawling now for haddock with another vessel. They are doing OK, but the quotas are now a third of what they were seven years ago. There is a ban on fishing for cod in the Irish Sea, and scientists say drastic reductions are needed to save the haddock. Increasing amounts of what fishermen land is illegal. Someone Tommy knows just landed 1,000 boxes of haddock, all illegal. Skippers keep no records and sell without documentation – probably for less than the going rate – to keep their cash flow and pay the mortgage on the boat.

Tommy is philosophical about everything except the prices he is getting for his fish. 'The scientists say there are no fish, but you never see a scientist in the Clyde when the cod are there in March and April. Come those months you will find thousands of boxes of cod out there.' There are times when some fisheries do well: this is a time when his is not. The pelagic boys who go out for mackerel and herring are doing well on rising quotas at the moment. What gets Tommy is when he does his job well and he and his partners catch hundreds of boxes of precious, legal haddock: even then, ungutted fish fetch only £20 a box when they should fetch £40, and gutted haddock fetch £30 when they should fetch £60. Tommy blames the price on imports from the Faroes and Iceland, where there is plenty of fish.

I haven't the heart to tell him my theory, which is that the domestic market is down because so many of his fellow fishermen are cheating – catching far more than their quota of what scientists now regard as endangered stocks of fish.

*   *   *

There is a woodcut on the wall of a fish-processing firm in Plymouth that I study as I wait to meet the owner. It features salty, idealised fishermen in sou'westers hauling a hemp net loaded with fat fish into an open sailing boat. It could be the Sea of Galilee at the time of Christ. It occurs to me that there is something timelessly right and something annoyingly wrong about this stylised image of fishermen. I struggle to define what annoys me about this picture. I come up with this. The picture tells the old truth that the real price of fish, the price you don't pay over the counter, has always been reckoned in men's lives. On average over ten years a British fishing vessel has been lost at sea every 12.5 days. There is a page of wrecks in every month's *Fishing News International*. But now the real price of fish has to be reckoned in a more complex equation: danger to fishermen versus damage to populations and ecosystems on which men ultimately depend. The world has changed. This is what reaching the end of the line for so many fish stocks means.

The depiction of strong men risking their lives as they battle against the elements isn't wrong, but the odds on the survival of the hunters have increased vastly since the days of steam trawlers, semaphore and 19th-century trawl captains, who made cabin boys trans-ship the catch at sea in open rowing boats. Safety increases by the decade, and who would not wish it to? Fishing is still an unenviably dangerous business in poor working conditions, especially in bad weather, in autumn and spring gales, in tropical storms and aboard long-distance fleets that fish near the poles. The dangers of fishing are known to be real, and fishermen relentlessly trade on this to their advantage. An old friend, a civil servant and press officer, once told me after bruising annual negotiations with Scottish fishermen and TV coverage that gets ministers hopping mad: 'Just when we've got those buggers where we want them, someone starts playing "For Those in Peril on the Sea".'

The risks to fishermen remain real, but our automatic sympathy for them has been modified in our more complex age by a new truth that has yet to find equilibrium: the odds on the survival of the hunted have been massively reduced. The development of fishing technology means that no area of the sea is inaccessible to nets or long lines. It used to be said of good Icelandic trawler skippers that they thought like fish. Now they can see like them as well.

* * *

The World Fishing Exhibition, Vigo, Spain, 2003. Advertising slogans for the latest developments in engines, boats, nets, lines, satellite phones, GPS plotters, sonar screens and computer systems for analysing what all of these are doing compete for the attention. Visitors to these several hectares of covered exhibition floor just by the airport are urged to 'Get the ultimate weapon on board' or to buy 'The world's largest trawl net' or to keep control of an entire fleet of satellite-connected, fish-tracking buoys through a 'Mobile Earth Station'. A hit this year is the New Zealand stand advertising the sea-mapping software Piscatus (slogan, 'Fish Hate Us'). Selling technology to fishermen is a macho business, rather like selling pesticides to farmers in the boom years of the 1970s, when top brands were given names such as Rapier, Impact and Commando. Where pesticide manufacturers tended towards images of combat to promote their products, fishing copywriters seem to prefer the language of science fiction or wars of mass destruction. I wonder why.

Legal, above-board public companies sell technology to fishermen, not all of whom will use it legally. Every major fishing gizmo supplier, boat-builder and some of the big players in fishing have decided to be seen during this three-yearly festival, which draws in sellers and buyers from over 30 countries in Europe, Africa and Latin America. Some, such as the giant Spanish company Pescanova, aren't really here to sell anything other than themselves. They are here to polish up their global image. Pescanova has a huge stand with hardly anybody on it, but video displays instead. Galicia's politicians queue up to speak at this exhibition, and an amiable 'dinosaur' from the Franco era, Manuel Fraga, who is in his 80s, speaks at the opening. The Spanish fisheries department has needed no encouragement to take a stand; amusingly, it confirms many people's impression of their countrymen by sporting posters in Spanish encouraging fishermen not to land undersized fish. Even the European Commission has been induced to take a stand, a sign of Spain's mysterious clout with the fisheries directorate general in Brussels.

Star of the show, in terms of the attention it receives, is the plotter, a flat-panel, high-resolution computer screen, perhaps 45 cm (18 inches) square, now the centrepiece of any large freezer-trawler's bridge. Most trawlers will have two – this year's model and last year's model – in case one goes on the blink. Fishermen don't use charts any more – not at this level anyway. The plotter brings together navigational data from all the onboard equipment on to one screen in front of the skipper's chair, with the contents of every other screen on the bridge just a mouse click away. Several versions are on sale by MaxSea (France), Simrad (Norway) and Furuno (Japan), together with screens for every other conceivable purpose. There are developments of the plotter, depending on what you are fishing for. For tuna purse-seiners and other pelagic boats, the MaxSea salesman tells me, the latest thing is to integrate navigational information with sea temperature contours from satellite weather channels. This enables the skipper of the purse-seiner, a sizeable ship that might carry its own helicopter or speedboat, to find the thermocline, the abrupt edge between masses of hot water and cooler water where the dynamism of the sea is greatest. Here plankton is generated and prey fish congregate, to be picked off by milling whirlwinds of tuna.

These feeding frenzies, in rainbow colours on the screen, are picked out in simulations run on the latest forward-facing, low-frequency sonars in the exhibition hall. One shows a feeding frenzy of tuna 4 km (2½ miles) ahead. Downward-facing, split-beam sonars, represented by other screens, show what is happening in the sea below. Tuna fishermen also use banks of bird radar, on sale here too, which are tuned to a pitch more sensitive than those used to detect aircraft, in order to pick up the flocks of birds that mark tuna shoals. Powerful, space-age pier binoculars, four or five times the size of those used by U-boat captains, are mounted on the bridge or in the crow's nests that many of these vessels have. These are used to spot the free swimming shoals that the purse-seiner's speedboat will race to enclose within the net.

Tuna fishermen know that tuna congregate around floating objects, having observed them around rafts or logs. At first they set their nets around these rafts to catch fish. Now each boat makes its own fish aggregation devices (FADs) – dozens of them, or as many as they can monitor. The FADs can be equipped with an echo sounder to monitor fish activity. They also monitor water temperature and speed of drift to assess whether the buoy has reached a thermocline. Each boat launches its FADs into the ocean currents, which in the Indian Ocean begin around Madagascar, and recovers them about 2,400 km (1,500 miles) to the northeast, around the Chagos Archipelago. FADs used to have radio beacons, and many still do because they are cheap, but the falling price of satellite telephone technology means that the latest gizmo to have is a buoy linked to the Inmarsat satellite. This sends a signal to a control centre that is capable of monitoring data from up to 600 FADs. If fishermen can afford a dozen or so buoys at €180,000 (£125,000) per buoy, it goes without saying that successful tuna purse-seiners are extremely profitable.

When it comes to trawler technology, the latest development in an industry that progresses as fast as the software industry itself is sea-bed mapping software, such as Piscatus 3D. This combines modern computer technology with the traditional echo-sounder to extract even more information from the sounds it sends to and receives from the sea bed. The result is that the fisherman can see into the depths in virtual reality. So that's why fish hate them. The information pack from Piscatus explains: 'We have developed a comprehensive 3-dimensional fishing tool that shows you exactly what is happening as you fish. See your boat, the sea bed, the fish, your gear: and even door-spread [the distance between the planing 'doors' holding the net open] in a real-time, animated 3D landscape.' It's a cross between *Star Wars* and *Flight Simulator*. This is the kind of stuff that fishermen buy when they are operating in waters that haven't yet been over-exploited and where fat profits are still to be made if they get there first. Equipment like this makes most trawl fleets in domestic waters look like cottage industries.

All this innovation can be puzzling at times to the tattooed, weatherbeaten men who come to exhibitions such as these. But the software people are not going to let puzzlement stand in the way of profit. 'The only ones frightened by our technology are the fish,' says the brochure put together by Piscatus's parent company, Seabed Mapping. 'As a skipper . . . you want to spend your time catching fish and not using computers. Seabed Mapping is part of the fishing industry and has developed Piscatus 3D alongside world-leading skippers. As a result, it's both highly effective and extremely easy to use . . . Being out of favour with the fish and making more money has never been this easy.'

MaxSea's personal bathymetric generator (PBG) does something very similar, allowing fishermen to create much more accurate 3D maps of a sea bed. Piscatus says its mapping software allows you to 'let the net drive the boat'. You can even fly your net down the line of a previous tow. Fisheries experts have spent decades debating how to quantify what they call 'technological creep', the ability of a single boat to catch more each year because of improvements in technology. What we are looking at here looks more like technological gallop.

For trawlermen fishing in deep water in the North Atlantic, the Indian Ocean, the Tasman Sea or the Pacific, underwater mountain tops, known as 'sea mounts' (see page 78), can be a particular hazard to nets. Consulting a 3D picture of an area of sea bed before and after MaxSea's PBG has been over it is the difference between looking at gently undulating prairie and the Rockies – the latter being an accurate description of the pinnacles and chasms on a rocky sea bed. The publicity material explains:

> The great advantage . . . is that it enables you to explore areas that would otherwise be too challenging – where other fishermen will not venture. Areas that have not been overfished and so promise particularly abundant catches. . . The unmatched accuracy . . . means you can go hunting for those 'hidden hideaways' (shelves, canyons and crevices) hitherto invisible and so virtually never fished.

A glowing endorsement follows from Michel Derosière, skipper of the *Fils de la Mer*. He says:

It's as if the water had been drained away and I can look right down and see exactly what the sea bed looks like. In the beginning I only got a fairly approximate view of the Channel floor. But now . . . it's fantastic. It means I can update the data in real time – so we can fish areas we used to avoid, whatever there might be down there, even the most treacherous shelves or rock formations. And we always know exactly where we are – to within a metre. This is a great new tool which soon pays for itself because there aren't many others fishing where we go, so we can be sure of hauling in much bigger catches.

One shudders to think that he's probably talking about the English Channel. This is fishing technology outstripping the speed at which most people can comprehend what fishermen are actually doing, let alone work out whether they approve of it.

Rather like a car salesman who lures the punter into the show-room with the latest Bentley or Mercedes parked out front, the MaxSea salesman points out that a lot of his company's 14,000 customers are skippers of small boats, and he has a pitch for them, too. He senses I'm probably not going to buy the top of the range PBD, the 'ultimate fishing aid'. He starts a different tack. 'You can map a big rock. You can go to 1.5-metres [5-ft] resolution. If you have lost your gears, you can get them back.'

The extent to which computer-based information technology has improved the fisherman's chances was described to me at a conference about developments in the deep sea held in conjunction with the exhibition. Halli Stefanson, an Icelandic fisherman who emigrated to New Zealand and now skippers a 2,500-tonne, 85-metre (280-ft) ship catching orange roughy, explained that these deep-water fish tend to congregate in a cloud on the tops of sea mounts. The process of catching them, with the latest com-puterised trawl-monitoring equipment, works like this: 'You can just drive the net on to the hill. It is a revolution for us. You try to drop the net as close to the top of the hill as possible and drop down the side. If you are lucky, you get about five minutes' fishing. This is typical of the orange roughy fishery. But you can get 17 tons as a result of two minutes' fishing.'

Acoustic net monitors, known as 'suitcases', show where the net is. Catch indicators, known as 'eggs', show when the net has 6, 10, 40 or 60 tons of fish in it. Net monitors are useful on any net, but they are virtually essential when managing one of the giant Gloria mid-water trawls, the biggest nets in the world, used to catch redfish on the mid-Atlantic ridges. The redfish is a solitary swimmer and does not shoal, hence the need for a huge net. The latest, on sale in Vigo, has a mouth opening of 35,800 square metres (43,000 square yards), large enough to catch half a dozen 747s or more flying in formation. These nets are six times larger in the opening than the original Gloria trawls of ten years ago, but this does not mean fishermen want to fill them up before they haul. Too many fish means mangled fish. A skipper in Reykjavik told me that he uses the net indicators to haul when there is a maximum of 6 tons in the net so that the fish are top quality and very fresh. If you go over that amount, the fish tend to get damaged and aren't worth as much. Alaskan pollock fishermen tell me they use net indicators for the same reason. Their fishery, which is better managed than many, prizes quality before quantity.

The potential of technology is, of course, neutral. It can be wasteful or clinical, depending on how it is used. It can be the tool of the policeman or the poacher. 'Blue boxes', which are satellite transceivers that transmit a position several times a day to tell regulators where fishing vessels are situated, are now required under EU law and in many other jurisdictions. They are, for example, required for tuna boats fishing under licence in Kenyan or Madagascan waters. The only trouble, I was told by Luiz Diaz del Rio, the director-general of Satlink, a firm that makes all sorts of satellite links, is that Spain, Portugal and Britain all require their vessels to have different sorts of blue box. The Spanish one, surprisingly, sends more information than the British. Some boats have all three if they are planning on spending time in all three countries' jurisdiction.

There is even a technical possibility of managing trawling to make it less indiscriminate. Fisherman could sense the size of the fish they are going to catch before they catch them, or set up a trawl to skim the bottom, causing minimal damage to the sea bed. It goes without saying, however, that it tends only to be the developments that increase the catch that are used by fishermen because there is no incentive to do otherwise. Here is Halli Stefanson again on the revolution he has seen in guiding the net over the past five years:

> This is all well and good, but recent information shows that sea mounts are very vulnerable to trawls. We must be careful how much we take out. Unfortunately, the mindset of many people in the industry has not changed since pre-technology days. Most sea mounts only yield fish for a year or two in any reasonable quantity. Orange roughy rely on a certain amount of biomass being present before spawning takes place. Bottom trawling can be disrupting the spawn for 3–4 years. In Iceland, where conservation practices have been in place since the 1860s, we fish on a sustainable basis. If we are to fish on sea mounts, we need to change our attitudes. Unfortunately, I'm not that optimistic. We had an agreement to protect sea mounts on the high seas in 1999 but it didn't work. Unscrupulous individuals exploited the loopholes. They made a lot of money, but the fishery has never been the same since.

In another session at the conference Grimur Valdimarsson, its Icelandic chairman, delivered a blunt conclusion to the deep-sea debate on fishing techniques. 'Technology,' he said, 'now makes it possible for us to catch whatever fish we like.' Malcolm Clarke, a fisheries scientist with one of the privatised fisheries bodies that now assess orange roughy stocks for New Zealand's companies, put it another way. He said: 'Our understanding of how to exploit the resource has moved much faster than our ability to manage it.'

CHAPTER 6

THE LAST FRONTIER

The oceans are the planet's last great living wilderness, man's only remaining frontier on Earth, and perhaps his last chance to prove himself a rational species.

John L. Culliney

THE FISH with which Europeans are most familiar come from the shallow seas of the Continental shelf or the surface waters of the open ocean. But there is plenty more water out there with fish in it within Europe's 200 mile limit and plenty more in international waters before you get to the abyssal depths. As fish stocks began to decline in European waters, enterprising fishermen began to look west. Out there was something called the Atlantic frontier, already being explored by the oil and gas industry. The technology to fish these depths is relatively new, part of the rapid development of boats, engines, winches and electronics that has happened in the past 30 years. The fish of deep water – the constantly dark world is generally agreed to begin 400 metres (1300 ft) down – are different in many ways to shallow-water fish and the regulations governing how fishermen exploit them are in their infancy within 200 mile limits and non-existent elsewhere. The concerns about their exploitation are more global but just as troubling as concerns about the use of shallower seas.

Before we descend to the dark world of true deep-sea fish, we must not overlook the unfortunate species, such as ling, tusk, Greenland halibut and blue whiting – once regarded by fishermen as deep-water fish – that no one seems to want in their scientific categories or management regimes. These are the fish of the continental slopes and mid-oceanic banks, which tend to find themselves just on the edge of countries' 320-km (200-mile) limits. As such, they are a cause of serious diplomatic trouble. Greenland halibut, or turbot, was the source of a major dispute between the EU and Canada over the international waters off the Grand Banks in 1995, and complaints about overfishing by EU vessels flicker on to this day. But now the hot issue is blue whiting.

Blue whiting, a member of the cod family, is found in waters from the surface down to 1,000 metres (3,300 ft), but is most often caught at 200–400 metres (660–1,300 ft). It reaches sexual maturity between two and seven years old, and can live for 20 years. Blue whiting ranges from the Barents Sea north of Finland to the mid-Atlantic around Iceland, and southwards as far as North Africa. In the west Atlantic it is caught in southern Canada and along the northeastern coast of the United States. It is a cousin of the whiting, but lives in shallower water and behaves like a pelagic fish of the deep sea, living near the bottom by day and making daily vertical migrations to the surface at night.

Surprisingly, for a fish that few people in Europe have heard of, blue whiting is about 50 times more numerous than the remnant North Sea cod, and nearly as numerous as Alaskan pollock, the largest remaining source of palatable fish in the world. So blue whiting is precisely the sort of fish that fishermen turn to when more accessible stocks run out. It is sold fresh and frozen, and is also processed into oil and fish-meal. I have never eaten it or seen it sold, but there is at least one Icelandic chef who says that it is an excellent table fish and has developed recipes in an attempt to persuade more Icelanders to eat it. Something or somebody, however, is eating blue whiting in large quantities because catches of it have trebled over the past seven years. This, I am told, has arisen partly from market developments in Russia and the Baltic states, and mostly because blue whiting makes excellent fish-meal for aquaculture. A gold rush is in progress. Catches of blue whiting off Iceland, the Faroes, Scotland and Ireland have risen from an already impressive 664,837 tons in 1987 to a staggering 1,554,995 tons in 2002. The worst culprit is Norway, which is catching nearly 800,000 tons a year – a catch it knows is unsustainable. The International Council for the Exploration of the Sea (ICES) in Copenhagen has said that a sustainable catch is more like 650,000 tons. Vast trawlers are using giant Gloria trawls at depths of around 500 metres (1,650 ft) to catch vast quantities of blue whiting, and are steaming back to port with the waves breaking across their decks because they are so full. Meanwhile, the politicians of the EU, Norway, Iceland and the Faroes sit on their hands.

I asked Kjartan Hoydal, secretary of the Northeast Atlantic Fisheries Commission, which attempts to control fishing in international waters off the British Isles, the Faroes and Iceland, what seemed to be the problem. He replied: 'With blue whiting it is obvious what should be done, but nobody seems to want to do it.' The EU, the Faroe Islands, Iceland and Norway have agreed a management plan, but no one could agree a share of the catch. This was all the more surprising, said Hoydal, because ample data existed on which countries caught what in earlier years, and it would be easy to base a quota system on it. Somewhere there was a political blockage. Meanwhile, countries set their own quotas, and fishermen took advantage of the chaos by mining down the stock.

'The last stock assessment from ICES was very good. But it is a short-lived stock and there are problems with assessing it. The fishery could be one or two year-classes [years when spawning survival was good] and that's it,' observed Hoydal bleakly.

* * *

The prognosis for the fish stocks of the deep sea, that is the dark depths from 400–3,000 metres (1,300–10,000 ft) where fishing is now feasible, are, if anything, bleaker. Although we know that deep-sea fish have been caught in quantity for more than 20 years, we are otherwise rather ignorant about them. We do know that in the entirely dark world below 1,000 metres (3,300 ft), fish tend to long-lived, less numerous and slow growing. They are either scavengers, waiting for infrequent 'food parcels' from the brighter regions above, or, like the many species of deep-water shark, predators that eat the scavengers.

The overall abundance of deep-water fish is much less than the fish of shelf waters, and catching them would be a costly, uneconomic business if it were not for their tendency to congregate around geographical features, such as raised banks and ridges, for feeding and spawning. Most dynamic and productive of these deep-sea habitats, but little known, are sea mounts, former volcanoes that rise off the ocean floor. Where sea mounts break the surface, such as in the Azores and Hawaii, we know them already as mid-oceanic islands. There are estimated to be 30,000 sea mounts in the Pacific, but only 6,000 in the Atlantic, and only 1,000 have been named worldwide. Far fewer have been studied. As George Clement, managing director of Seabed Mapping, the New Zealand company that manufactures the Piscatus software that compiles 3D maps of the sea bed, put it: 'The sea floor is less well mapped than the surface of the moon.'

The strong localised currents and up-wellings created by sea mounts make them hotspots for plankton, and consequently for small fish and the large fish that prey on them. Scientists tell us that sea mounts, oases of diversity in vast expanses of open ocean, are likely to support large numbers of undiscovered species. They certainly support communities of suspension feeders, corals, sponges and sea fans that filter organic matter from the water rushing over the summit. Orange roughy, among other fish that gather to spawn on the top of sea mounts, feed on prawns, squid and small fish that float by. Further down a sea mount the coral becomes less dense, and lobsters and sea spiders hide among the rocks. Unfortunately, the opportunity to study even a represent-ative selection of the world's sea mounts before trawling fundamentally alters them is fast running out.

While sea mounts are the most spectacular habitats, there is enough diversity in the species that live in the micro-climates caused by ridges and currents on the continental slopes of the Atlantic basin to keep a biologist busy for a lifetime. Dr John Gordon is an honorary fellow at the Dunstaffnage marine laboratory near Oban on the west coast of Scotland, where he worked as a biologist until his retirement. Now he just works there anyway, mostly unpaid. He began to study deep-water, bottom-living fish when this was regarded as an extremely esoteric subject. He saw his first catch of deep-water fish on the research ship *Challenger*'s voyage to the Rockall Trough in 1973. This area is now the centre of intense fishing activity by French and Spanish trawlers, which were the first to specialise in the deep sea as shelf seas became overfished.

Dr Gordon now finds himself a lone voice in warning that the fisheries of the Rockall Trough and Hatton Bank are being subjected to unsustainable fishing pressure – in other words, they are fast being mined out. In 1997 a new British government decided to ratify the UN Convention on the Law of the Sea. It therefore renounced its claim to a 320-km (200-mile) limit around Rockall, a granite outcrop about 25 metres (80 ft) high and 320 km (200 miles) to the west of the outer Hebrides. The reason given was that international law now views claims based on uninhabited rocks as untenable. However, this does not prevent other countries continuing to inhabit them occasionally, and defending *their* claims. The consequence of this decision for fisheries does not seem to have been considered. It means that much of the Rockall Trough, an area of deep water that has been studied since the 1860s, and the whole of the Hatton Bank are now in international waters. As a result, they have been the subject of a fishing free-for-all.

At the conference in Vigo, where Halli Stefanson told us about the excitement of fishing the deep sea with the latest technology, Dr Gordon explained the differences between the fish of shallow seas and the weird and little understood animals of the deep. Between 400 and 2000 metres (1,300 and 6,600 ft) down, he told us, there are more than 100 species of bottom-living fish. Deep-water fish have either deciduous scales or none at all, making them more vulnerable to damage by trawls, even if they escape through the mesh. Deep-water fish, such as orange roughy, scabbard fish, smootheads, rabbit fish, scorpion fish and grenadiers, tend to have large heads and tapering bodies, so undersized fish get caught in nets that would allow the young of differently shaped shallow-water fish to escape. There is no wave action at depth, and such currents as there are do not generally have the same force as those in shallower seas, so lost or snagged fishing nets tend to go on fishing, often for decades, as 'ghost nets'.

The other reason why fishing has a greater impact upon deep-sea fish is age: some of these fish are remarkably long-lived. The orange roughy, for example, is now assumed by most scientists, including Dr Gordon, to live up to 150 years. It takes 20 years to mature and does not reproduce until it is about 30. (It must be noted that there is still some controversy about dating old specimens of orange roughy. Based on growth rings found in otoliths – fishes' ear bones – like those found in trees, dating assumes that one ring equals one year, which would make some of the roughy that have been caught very old indeed. Some scientists and fishermen, however, are convinced that the rings do not represent years and that the top age of a roughy will turn out to be about 18.) Even when it does mature, the orange roughy is less productive than shallow-water species. It lays only tens of thousands of eggs, an order of magnitude less than the cod's several million.

A similar pattern of long life and low fecundity is found in other deep-water fish. Round-nosed grenadiers mature at 8–10 years old and live to 75 years. Smootheads, such as Baird's smoothead, live to 38 years, and the leafscale gulpher shark is known to live to 70 years old and to have only 6–11 live young. Large numbers of these fish are known to be caught and discarded by long-line vessels targeting other species, but few records are kept. The effect of fishing on these non-target species, just like their basic biology, is almost completely unknown. Dr Gordon was surprised when he went aboard research ships carrying remotely operated survey vessels (ROVs) to find that the images they beamed up most frequently from the depths were of fish that did not get trapped by trawls at all. The most numerous was the cut-throat eel (*Synaphobranchus kaupi*), which is thin enough to survive being caught in all but the most fine-meshed nets. The cut-throat eel is the most abundant species in the Rockall Trough and the Porcupine Sea Bight, off southwest Ireland. Dr Gordon found it was also in the Bay of Biscay when he was invited on to the French vessel *L'Atalante* in 1992.

It is inherently surprising that anyone wants to eat a fish like orange roughy that, as far as we can work out, outlives even the longest-living humans. It is known as the Queen Mother of fish. Just as remarkable is the fact that the main European market for this fish, which produces firm, bland but relatively tasteless white fillets, is France, a nation renowned for its demanding culinary standards. France began catching blue ling in the 1970s, and started on orange roughy in the North Atlantic 1991. It is now the leading country in Europe for catching and eating deep-water fish. How did this come about?

The unattractiveness of many deep-sea fish was overcome by filleting them, which makes them look like any other white fish fillets. Then, French marketing people tackled the problem of unfamiliar names, and here they succeeded beyond their wildest expectations. Abandoning the dull common or scientific names the fish were lumbered with, the marketing people equipped them with dashing, military-sounding names from a glorious Napoleonic past, designed to sound a note of pride in French breasts. Black scabbard fish became *sabre*, orange roughy *empereur* (which happens to resemble the name for swordfish in Spanish), and round-nosed grenadier became simply *grenadier*. At the time the new species were reaching the market in the early 1990s, the processing industry was crying out for more white fish fillets, so once the marketers had solved the naming problem, the niche was filled. Orange roughy also finds ready markets along the west coast of the United States, where its meaty fillets are sold as the fish for people who don't really like fish.

The effect of trawling in deep water in the North Atlantic has been to reduce all known fish populations there to around 20 per cent of what they were in the 1970s. That was true of predator species, such as round-nosed grenadier, black scabbard fish and sikis (a kind of deep-water shark). With orange roughy, the attrition appears to have been worse. The fishery in British waters started in 1991. By 1994 the catch rate was 25 per cent of initial catch rates. Effectively, the known orange roughy populations of Hatton Bank, Porcupine Sea Bight and Rockall Trough have been mined out within a few decades of being found.

Stocks of round-nosed grenadier west of the British Isles are now so low that they are below the precautionary level set by the ICES – the level at which fishing should theoretically stop. For most deep-water species a sustainable exploitation level is around 2 per cent of the original stock, compared with 20–30 per cent for shallow-water fish. Scientists say it may simply not be possible to 'harvest' a deep-sea stock sustainably – that is, take a commercially viable number of fish every year. All that we can perhaps do is harvest them down to a certain known level from which we hope they will recover, then leave them alone. This assumes we know enough about the biology of the fish to know the level at which the population can recover and can measure it. It also assumes we have control over the resource, which is doubtful considering the international waters are managed by the Northwest Atlantic Fisheries Commission, which controls the Hatton Bank. The gloomiest conclusion came in the mid-1990s from Phil Aikman, author of a Greenpeace report on deep-sea stocks. He said we should not be exploiting deep-water fish at all.

The one country that claims it has finally found a way of fishing orange roughy sustainably is New Zealand. Its exclusive economic zone (EEZ) is mostly deep water, which includes extensive ridges and sea mounts, so it got into deep-water fishing 20 years ago. There was a time in the 1980s when New Zealand fishermen were catching 100,000 tons a year of orange roughy, a significant proportion of the world catch. Then scientists found that stocks were a fraction of the size they thought they were, so fishing had to be scaled back. The difficulty was in estimating the stock size. Scientists – who in New Zealand work for private companies and are hired by fishermen who pay the cost of research – differ in their opinion of the stocks even now by a factor of three. Nine out of ten roughy stocks are not looking good. Some are closed to fishing altogether, and only one appears to be rebuilding, but even that could be because the form of scientific assessment has changed. Malcolm Clark, who works for a privatised fishing science company, told the Vigo conference: 'Orange roughy has an uphill battle when it comes to sustainability. Our overall experience in New Zealand doesn't look good. A lot of management lessons have been learnt.'

George Clement, who is not only managing director of Seabed Mapping but also of the Orange Roughy Company, which manages the fishery, says the orange roughy is now being fished sustainably after many years of scientists being over-optimistic about how many fish were there. Clement, a believer in the theory that the orange roughy reproduces long before it is 30 and lives to nothing like 100 years old, believes some stocks are actually rebuilding. Unfortunately, it will be ten years or more before we find out if he is right – by which time a lot of more mature roughy will have been caught.

In Europe, there doesn't seem to have been the same over-assessment of stocks that happened in New Zealand. Instead, there has been bureaucratic inertia in the face of a whole new area of the sea to research and control. Repeated warnings have been sounded, but the European Commission has reacted far too late. In 2000 it announced quotas for deep-sea stocks, but these are far larger than recommended by ICES. The best summary of the situation in the waters off the British Isles now being fished mostly by France and Spain is provided by Dr Gordon:

> There is general agreement among scientists, the fishing industry and the politicians that the fragile, deep-water stocks are seriously over-exploited, but political imperatives dictate that uncertainties and inconsistencies in the scientific assessment and advice are used to postpone the urgent action that is required. It is perhaps not much of a consolation, but

at least in the Rockall Trough, we know a lot about the ecosystem that is being destroyed, while in other areas, such as the Hatton Bank, we will never know what is being destroyed.

He's right. To our children and grandchildren it will sound like slim consolation.

CHAPTER 7

THE INEXHAUSTIBLE SEA?

LOWESTOFT, ENGLAND. The irony slaps you in the face like the January gale. The fish dock is all but empty, the herring drifters gone, the last beam-trawlers that sailed from here broken up or sold to flail the bed of some other ocean, but the government laboratory that was set up to ensure the survival of a plentiful supply of fish lives on, dominating the town, costing taxpayers money, monitoring a sea that it was supposed to save. The laboratory's new 73-metre (240-ft) research vessel, the *Endeavour*, which cost £24 million, dwarfs other boats in the dock. Yet, despite all this public expenditure we have a sea where two-thirds of the major commercial fish species are 'outside safe biological limits'. That is the accepted scientific way of saying that they are close to the point of no return.

The central problem for fisheries science is counting fish. You cannot, even with current technology, see how many fish there are in the sea, so you have to find ways of counting them. What happened? Have scientists around the North Sea got their numbers wrong? Or are the endeavours of scientists routinely overwhelmed by those of politicians and fishermen with different agendas? Or are scientists effectively just another interest group, with only a marginal interest in achieving the conservation of fish on their path to career preferment? Such explanations have all been offered at different times and on different continents for the manifest failure of modern science-based management, with all but a few exceptions, to conserve fish.

Not being a scientist, I find myself generally unenlightened and annoyed by the vast amount of bureaucratic language churned out by the majority of fisheries scientists. They might believe they are being neutral and non-judgemental, but they are in fact betraying a clear bias in favour of large-scale industrial fisheries as the principal use of the sea. To avoid being shaped by their patterns in the sand, I will tell the story in a different way, as history, through the lives and perspectives of a few individuals.

The Lowestoft fisheries laboratory is housed in a red-brick building, formerly the Grand Hotel. Today it is part of a government agency with the forgettable name of the Centre for Environment, Fisheries and Aquaculture Science (CEFAS), but it is popularly known as the Lowestoft fisheries lab. Its deputy director and senior scientist, Dr Joe Horwood, is a cautious man whom one detects is temperamentally averse to making strong statements. Yet it has been his job to issue regular and unambiguous warnings to fishermen and politicians on the state of the dwindling shoals of cod, haddock, plaice and sole in the North Sea for nearly a decade – warnings that he has come to expect will be largely ignored. It is a challenge he rises to without enthusiasm. Wearing a lugubrious expression, Dr Horwood shows me a corridor lined with photographs of his predecessors. Like him, nearly all are careful-looking men, government scientists ready with a cautiously chosen word for the minister's ear, presented in the detached but pliant manner of the British civil service. This implies that the minister can do what he likes with the information, but what he is hearing is the considered opinion of the Lowestoft laboratory.

One figure stands out in the portrait gallery. It is the only man – they are all men – who is not wearing a jacket and tie and sitting at a desk. He is Michael Graham, director of fisheries research from 1945 to 1958. He is standing on the deck of a trawler dressed in a fisherman's sweater and apron, pipe in mouth. He is busy scaling a cod.

Michael Graham was a scientist whose contribution to the theory of fishing – which chiefly involves describing overfishing – was large and timely. The years 1945 to 1958, when he was director of the Lowestoft laboratory, saw it achieve world leadership in the understanding of sea fish populations. He was an idiosyncratic employer, ordering that young scientists, or naturalists, as he insisted they were called, went to sea.

To a later generation, trying to explain why the North Sea and other seas declined and their fishing ports collapsed, Graham is a seminal figure, both for his influence on the golden age of fisheries science and for his restless advocacy of 'rational' fishing and the conservation of biological systems on which humans depend for food. The starting point for anyone trying to understand his views and ideas on fishing is *The Fish Gate* (1943). In this slim book, in the muscular English of wartime, Graham explains why the idea of free fishing failed in his lifetime and why that failure was compounded by the scientific errors and evasions of previous generations.

When Michael Graham started work at the Ministry of Agriculture and Fisheries' Lowestoft laboratory in 1920 it had been a demonstrated fact for at least 35 years that industrial trawling can cause overfishing. In Britain, where fishing was first industrialised with steam trawlers working out of Hull, Grimsby and Aberdeen around the 1860s, it was the fishermen, not scientists, who first pointed out that fish populations in parts of the North Sea were being systematically wiped out. The absence of fish in their old grounds required them to steam further out to sea to find new ones. The scientific establishment, in the shape of Thomas Huxley, met this observation with contempt. Huxley chaired a commission in the late 1860s looking at whether fishermen had grounds for concern. He was still expressing the view that commission formed nearly 20 years later in 1883:

I believe that it may be affirmed with confidence that, in relation to our present modes of fishing, a number of the most important sea fisheries, such as the cod fishery, the herring fishery, and the mackerel fishery, are inexhaustible. And I base this conviction on two grounds, first, that the multitude of these fishes is so inconceivably great that the number we catch is relatively insignificant: and secondly, that the multitude of the destructive agencies at work upon them is so prodigious, that the destruction effected by the fishermen cannot sensibly increase the death-rate.

Another parliamentary inquiry of which an ailing Huxley was a member reversed those conclusions within the decade.

The decline in the fisheries progressed at an alarming rate over the next 25 years, and would have continued to do so, were it not for the First World War, when most trawlers were either requisitioned for other tasks or unable to put to sea because of mines or hostile shipping. This gave fish stocks a chance to recover. During the war, with little fishing possible, fish prices more than doubled. In 1914, the average daily landing from a trawler in the North Sea was 716 kg (14.6 cwt). When fishing resumed in 1919, the average daily landing was 1,542 kg (30.6 cwt). But within five years the boom had turned to bust. Fishermen, who had invested in new boats and gear, were catching too much. Prices collapsed, making them fish all the harder. Trawlermen had to sell boxes of 90-cm (3-ft) cod for fish-meal. Then stocks became depleted too. Parliamentary inquiries expressed increasing concern about North Sea fish stocks throughout the 1920s and 1930s.

Michael Graham's given task on joining the Lowestoft laboratory was to study the North Sea cod fishery. He continued, with breaks for field studies in Africa and Canada, throughout the 1920s and 1930s, describing the cod's life cycle and spawning grounds, and showing the age composition of the fishery through a laborious process of scale-reading. In a 1935 paper Graham conclusively showed that the stock was overfished. His book, *The Fish Gate*, reminds us that by 1939 the North Sea was exhausted and many fishermen unemployed – the inevitable result of free fishing. Graham distilled the observations of 20 years at sea into his 'Great Law of Fishing': *Fisheries that are unlimited become unprofitable.* Graham said this was a law demonstrable by experience, not scientific theory, an interesting distinction. As long as the effort is free and urged on by competition, fisheries will eventually fail. The Great Law, he pointed out, was only fruitful in its converse form: *that limiting the effort will restore profit to a fishery.* His next objective was to find a way of showing exactlyhow much to limit fisheries to provide the optimum catch.

When Graham returned to Lowestoft after the Second World War – which he had spent latterly as a scientific adviser to the RAF, appraising the efficiency of new aircraft and equipment – it was as director. Sidney Holt, one of the young mathematicians Graham hired straight from university, remembers it was a time of general optimism and determination that the post-war world would not repeat the mistakes of the past. Half the fishing vessels in Europe had been sunk, and while there were going to be new vessels, Graham believed Europe could avoid building up an excessive fleet, as it had done after the First World War, if it had a clear way of calculating the size of stocks and what fishing pressure would do to them. The job of a scientist was to improve matters, Graham told Holt in his first week at Lowestoft in 1947. The task was urgent: fish stocks had been given a six-year breather that had enabled them to recover from overfishing before the war. One of Holt's first tasks was to look into whether this suspected recovery had actually occurred, which it had. After several years' hard work, he and his colleague Ray Beverton produced their mathematical model of how plaice and haddock populations behave under fishing pressure. Their findings were eventually published in 1957 but long before that began to be internationally influential.

Beverton and Holt had produced the means by which single fish populations could be managed reasonably successfully – provided scientists fed the right data into the models and made appropriate assumptions about a range of things, including the natural and human-induced mortality of fish and provided politicians acted on these recommendations and limited fishing accordingly. These provisos became increasingly important as time went on.

Michael Graham believed that there must be an optimum point somewhere in the relationship between a sustainable population size and fishing intensity. Other scientists, such as M. B. Schaefer in the United States and W. E. Ricker in Canada, came to define this point later in the 1950s as maximum sustainable yield (MSY). There were some who, unlike Graham, believed that this was an objective to strive towards. Sidney Holt says this characterised the difference between European and American thought during the 1950s. Europeans such as Graham believed that fisheries should be managed for long term stability and to preserve jobs. Scientists in the United States, mindful that their constitution required any fishery to be open access, needed to make room for expansion. This is where trouble began.

The pursuit of maximum sustainable yield encouraged fishermen to drive down the original population to a lower level – taking up to half of the total spawning stock every year – in the belief that this would boost the productivity of the population. A smaller population, the theory went, would grow faster and reproduce earlier because it had a more plentiful food supply. The problem was simply that the MSY concept tested the Beverton-Holt single-stock model to destruction. To decide what level of catches approached the magic figure of MSY there was little room for error. Scientists needed accurate figures for fishing mortality (i.e. fisher-men must not cheat by misreporting catches or discards) and natural mortality. There were also the dangers of missing or misinterpreting environmental forces or unforeseen predator effects that existed outside their models. Complete accuracy is, of course, impossible, but without it the result is that catches will be allowed to rise too high. Result: overfishing. A near-definitive demolition of MSY as a concept was written in 1977 by the Canadian biologist Peter Larkin. His short poem on the subject is better known:

> Here lies the concept, MSY.
> It advocated yields too high,
> And didn't spell out how to slice the pie.
> We bury it with best of wishes.
> Especially on behalf of fishes.

Lay people may find it strange that the concept of fishing to provide maximum sustained yield, despite being comprehensively discredited, remains the unaltered objective of several international conventions, such as the one governing tuna stocks in the Atlantic, which was negotiated in the 1960s. Even more strangely, it was included in the part of the declaration of the 2002 Johannesburg summit on sustainable development that applies to fisheries. This said that to achieve sustainable fisheries the following actions were required 'at all levels': 'Maintain or restore stocks to levels that can produce the maximum sustainable yield with the aim of achieving these goals for depleted stocks on an urgent basis and *where possible* [my italics] no later than 2015.'

I was at the Johannesburg summit and remember noticing a flurry of activity until that *where possible* was inserted in the plan for the next two decades. Later on, at a social gathering, the assembled diplomats of the world congratulated themselves on a good job done. Some of us reflected that the self-congratulation was because they now didn't have to do anything at all.

* * *

Berlin, June 2003. A leafier, pleasanter city than I remembered. I ran into Sidney Holt at a meeting of the International Whaling Commission, which he still attends every year, though now in his seventies. We arranged to meet up for a leisurely breakfast around 8am in our vast, impersonal hotel. We were still talking when the waiters were clearing away the remains two and a half hours later. Holt said he had been looking again at the work of his early years and at the disaster (his word) that had happened to the fish of the North Sea in his absence. Here, it occurred to me, was the epilogue to the golden era of fisheries science being expounded in front of me, as Holt urgently drew dome-shaped yield curves on paper napkins.

Where did he think things went wrong? Thereby hangs a personal tale. Beverton stayed at Lowestoft until the 1970s, when he went off to run the Natural Environment Research Council.

Holt, more of a radical figure, who had been a firebrand communist while at university during the war, went to work for the UN Food and Agriculture Organisation in Rome in 1953. In the 1970s he began to work for environmental groups and individuals, lending his mathematical modelling skills to the battle to stop whaling.

Holt told me that he and Beverton differed on many things and they didn't see each other very often. They argued when they did meet about where the blame should lie for the increasingly parlous state of fisheries around the world. Beverton argued that overfishing had much to do with authorities failing to act appropriately and in good time on scientific advice. Holt argued – as do contemporary scientists, such as Daniel Pauly – that aspects of the models routinely used by biologists actually induce overfishing. Just before Beverton's death in 1995, he and Holt eventually agreed that it was a bit of both. In a lecture he sent to a conference in Vancouver that year, being too ill to deliver it himself, Beverton noted: 'There is a strong inverse association between the growth of fisheries science . . . and the effectiveness with which it is applied.' Kindly, he concluded: 'Events of recent years raise in my mind the lingering suspicion that Sidney [Holt] might have been right to the extent that we have been trying to be too clever and have failed to see the wood for the trees.'

A prime example of this, says Holt, was the assumption made by scientists for more than 30 years, supposedly based on his and Beverton's work, that the number of juveniles reaching fishable size each year has nothing to do with the number of parents in the sea. This would obviously be untrue for mammals such as dogs, cats, humans and whales. It would even be untrue for marine creatures, such as sharks, that have few young, so the number of parents is undoubtedly the most important factor in how many offspring fish produce. Fish such as cod, however, can produce upwards of 7 million eggs, of which only a tiny fraction survives. For many years it was therefore argued that there would always be enough adult fish to produce the usual number of young, and the main influence on the numbers of juvenile fish were environmental factors, such as temperature and predation. This argument was used to justify higher catches than might otherwise be the case. I remember being subjected to patronising expositions of this counter-intuitive – and, to an informed layman, unbelievable – argument by senior scientists, officials and even politicians in London and Brussels. All brushed aside the layman's observation that logic dictated that the population must at some time reach a level where the number of parents mattered. So what was the level that the number of parents did begin to matter? Was it a low number or a high one?

This, as it happens, is precisely Holt's point. He and Beverton found no detectable relationship between the numbers of plaice of catchable size recruited annually to the North Sea plaice stock in the late 1940s and the number of parent fish. But the North Sea plaice was at that time a large population that had just spent six years recovering from overfishing. He and Beverton knew then, he says, that this observation wouldn't always hold. 'Even in the 1950s, we knew it didn't and this could cause problems,' he said. In the event, it led to 'progressively more and more dangerously optimistic quotas'.

Another danger with population models is that they usually assume that a small fish population will be more productive than a large one. In nature, however, depleted populations do not behave like healthy ones. They are less productive. This phenomenon is called by a variety of names – depensation, the depopulation effect, or the Allee effect, after the ecologist W. C. Allee, who described it in insects during the 1930s. The Allee effect claimed the passenger pigeon, a creature now extinct, but once so numerous that it used to darken the skies over North America. The same effect kicked in with the Northern cod on the Grand Banks of Newfoundland, and is in danger of raising its head with cod in the North Sea. Alarmingly, the effect is not distinguishable from simple fishing pressure until it is much too late to do anything about it.

The dangerous dogma that the number of parents did not matter meant that scientists were more inclined to blame the failure of stocks to regenerate on environmental variability. There is continual, dynamic environmental variability in the sea, not all of which is regular, like the warm current pulse in the Pacific known as El Niño. This phenomenon was thought to be a factor in the 1970s' collapse of the Peruvian anchoveta, the largest fish stock in the world, and greenhouse warming may well be responsible for making the North Sea a less favourable place for cod today. But in these examples there is, or was, massive fishing pressure. Without fishing, which unquestionably kills millions of potential breeding adults, the population would have stood a far greater chance of adapting to environmental change.

Holt told me that rather than aiming high to stimulate fish populations to grow, we should be aiming low – for a catch that is well within the biological limitations of the species, the possible fluctuations of its environment and possible errors in estimation. That almost certainly implies lower catches than currently permitted in any commercial fishery in the world today. If one were to subject the world's fisheries to the Revised Management Procedure devised for controlling catches of whales, most of which are now too late for it to be used, most fisheries would be closed overnight.

Why some over-exploited stocks fail to recover while others, such as the anchoveta come back, is not well understood. Jeff Hutchings, a marine biologist at Dalhousie University in Canada, has analysed 90 stocks worldwide, many of which have experienced massive declines due to overfishing. Most, with the exception of the fast-growing Atlantic herring, showed little signs of recovery 15 years after their collapse. Hutchings told *Nature*: 'The life history of species matters. Small, early-maturing, mid-water species like herring might recover faster than late-maturing, bottom-living species such as cod.' Hutchings's view was that if stocks were not recoverable, a far greater degree of precaution was needed in setting catch limits in the first place. 'If there is not much we can do after the damage is done, then it is an even stronger case that we should not let fish stocks fall below safe levels.' Estimates of sustainable catches have become much more conservative since the 1970s, and more so since the Grand Banks disaster, but almost nowhere, with the arguable exception of Iceland, and for a number of pelagic species, have catch limits been set low enough for populations to rebuild.

The Grand Banks is the textbook case of failure in fisheries science. An army of scientists in one of the world's richest and most advanced nations managed to destroy one of the richest fisheries in the world, while convincing themselves for a decade that they were doing no such thing. The Newfoundland cod collapse was the nightmare that shook the world out of its complacent assumption that the sea's resources were renewable and being managed in an enlightened manner.

It is not possible to say that with better science Canada would have avoided disaster, because there is no way of predicting what mutton-headed decisions politicians would have made if scientists had noticed earlier that the cod was in trouble. Politicians in Canada have a long record of using any kind of scientific uncertainty to argue for higher catches. What one can say is that according to the models scientists were using in the early and mid-1980s, the Grand Banks should still be full of cod.

The great advantage of hindsight is that one becomes aware of the dissident voices prophesying disaster that were ignored by the bureaucratic machine that was Canada's Department of Fisheries and Oceans, and by its elite and secretive bunch of scientists responsible for assessing stocks. As far as one can see – and this area is fraught with scientific papers trying to write and rewrite history to excuse some and blame others – a succession of mistakes was made, which led to stock assessments being grossly inaccurate from the late 1970s. All the mistakes involved assumptions fed into population models, which proved the old computing adage: garbage in, garbage out.

The first, glaring, error was made in the late 1970s and early 1980s, when Canada declared jurisdiction over 320 km (200 miles) of the Grand Banks and ejected trawlers from the Soviet Union, Poland, Britain, Spain and other nations that had hammered cod stocks there to a tenth of their level in the 1960s. There was a political opportunity, between 1978 and 1983, when scientists could have set a more cautious course because the Canadian fleet did not have the capacity at that time to overfish the cod itself. But stock assessments of cod in the late 1970s failed to reveal the extent of destruction to the spawning stock, which was eventually revealed by scientists going back over the same figures a decade later.

What happened exactly was that Canada did not have credible data for some of the areas within its new fishing zone – because Germany were the last nation to have conducted surveys there when the Grand Banks were international waters. As a result, commercial catch per unit effort data were used to 'adjust' or 'tune' models that were supposed to be based on random trawl surveys. The assumption that the commercial catch data conveyed an accurate picture of stocks across the sea bed, rather than on the grounds where fishermen went to fish, turned out to be catastrophically mistaken. Where commercial catch data showed the stock had declined by 70 per cent, the stock had actually fallen by 90 per cent since 1962.

Convinced that cod stocks would now recover quickly to pre-1960s' levels, the provincial government of Newfoundland and the government of Canada called for long-term catch predictions – based on these erroneous assessments. Scientists duly predicted catches of 400,000 tons a year by 1990. These predictions now look incredible. Stocks never rose enough to allow catch limits higher than 260,000 tons. Initially the cod did stage a recovery, though never to the levels predicted. By the time it had been worked out, much later, that commercial catch data and trawler surveys were showing different trends, it was too late.

Now classic errors, identified by Sidney Holt, were made. The importance of a large, healthy spawning stock was ignored. The overfished population did not become productive: it turned out to be less productive than the previous larger one. By the early 1980s, on the basis of rosy long-term forecasts, heavy subsidies were pouring into the province to build Canada's own trawler fleet. Carl Walters of the University of British Columbia and Jean-Jacques Maguire of the DFO in Quebec published a paper in 1996 which concluded: 'Even if scientists had suddenly reversed their conclusions and called loudly and publicly for harvest restraints, by about 1982 an institutional juggernaut had been set in motion that no political decision-maker would have dared try to stop until it was too late. The window of opportunity had closed.'

Astonishingly, Canada's scientists continued to think that they were setting annual catch limits at 16 per cent of the fish, which in theory would allow stocks to increase rapidly. Later analysis would show that fishermen were catching more like 60 per cent of the adult fish each year. Warning signs began to appear: smaller fish, an indication that they stood less chance of surviving; the 'dragger' (trawler) fleet was fishing a smaller and smaller area of ocean; inshore fishermen complained, as they had since the early 1980s, that catches were going down. To compound the muddle, the technology on board draggers had increased as a result of subsidies, thereby increasing the catch per unit effort – one of the ways scientists measured the stock. But the Department of Fisheries and Oceans had no way of revising its estimates to take account of 'technological creep'.

Every autumn the DFO research vessel steered its random course across the banks, counting how many fish it caught. By 1989 it was showing large areas of empty ocean. But fishermen said there were still plenty of fish because their fish-finders were detecting hotspots where the dwindling shoals of cod were congregating. Schools of cod or haddock huddle together when they are depleted. Other fish, such as hake, do not.

The scientists couldn't decide what to do. The whole theoretical basis on which they had built their assessments was crumbling. They advised a catch limit of 125,000 tons, less than half the 266,000 tons of 1988. The fisheries minister of the time refused to agree to it for fear of upsetting fishermen. He set the quota at 235,000 tons. Lesley Harris, a former president of Memorial University in St John's, Newfoundland, said that the DFO should have insisted. 'But scientists being scientists, they weren't prepared to make absolute statements about anything.'

Many remember Jake Rice, chief of research at the DFO in St John's, cheerfully going on television before the collapse, saying that lots of young cod were coming through, several year-classes of them. These young cod never made it. Millions were caught and discarded in the last two years of the fishery by fishermen who refused to believe how serious things were. By June 1992 the DFO realised that there were no cod left old enough to spawn, and by then even the fishermen were showing concern. A year-long moratorium was declared and extended indefinitely in 1993.

The failure to be truly open with data and to go public about uncertainty is perhaps the greatest lesson for science from the Grand Banks disaster. With more transparency, the errors would have been detected earlier. Ransom Myers, then at the DFO, says he was not allowed access to the data used to calculate stock assessments as it was the preserve of a charmed circle he calls 'the tribe'. Members of this tribe, he believes, were unduly concerned with protecting the reputations of the assessors and their political masters, and this got in the way of accuracy. Why does he think a scientific team, supposedly paid to reward its independence and expertise, failed to spot such huge errors for so long? 'Every choice,' he says, 'was about "What is going to get me rewarded?" Positive news makes people happy. Bad news is not always appreciated.'

Professor John Shepherd of Southampton University, a former deputy director of the Lowestoft laboratory and one of the small group of fisheries scientists to emerge from the British government machine with integrity, once put it better than anyone else: 'We will always have a problem until we recognise that a scientist's first duty is to the truth. His second duty is to the public interest and his third duty is to the minister.'

What else does the Grand Banks crash tell us? Could such huge scientific errors happen again? The answer, according to Carl Walters and J. J. Maguire in 1996, is yes, because systematic over-estimation of stock size and inadequate attention to the spawning stock are common problems, difficult to expunge from single-stock models. Maguire became head of research at ICES and helped it to clean up its act by promoting transparency, cracking down on false assumptions in computer models and promoting a greater degree of precaution in deciding catch limits. Since then ICES has consistently recommended a total ban on catching cod in the North Sea. But many other scientific bodies, notably those responsible for tuna, are suspected by other scientists of still coming up with gung-ho stock assessments that bear the hallmarks of wishful thinking.

One positive development since the Grand Banks disaster is the number of fisheries scientists attracting funding from conservation bodies. Diversity of funding is good, for it overcomes the traditional scientists' problem that whatever the data says, the paymaster – generally a government – gets to call the shots when the summary is written. More open diversity of opinion makes fisheries scientists have to defend the catch limits they recommend not just to self-interested fishermen, who might wish them to be higher, but in the arena of world opinion against knowledgeable people from related disciplines, who may well think the quotas are far too high.

So what are we to conclude about Michael Graham and his colleagues in the 1940s and 1950s, who constructed theoretical tools that they hoped would enable the sea to be managed in a rational, benign way? Did they try as hard as they might or were their efforts thwarted? Cynics would say that the science of counting fish merely served to stretch out the period between boom and bust for 40 years, from the late 1950s to the late 1990s – in other words scientists simply measured decline without actually managing to reverse it when things got tough. Actually, the only accurate way of judging scientists – from Europe or anywhere else – is to ask whether, if politicians had taken their advice, the fish stocks of the northeast Atlantic would now be in a better state. The answer, unequivocally, is that there would be more fish.

On the eastern side of the Atlantic, scientists have made many errors, but the dire state of Europe's sea derives overwhelmingly from poor governance. If you look at the declining line of fish catches on the graph over the past 15 years, ICES got the trend right. You might rightly criticise it for failing at any time to recommend that quotas be slashed heavily to allow stocks to rebuild. To an outsider there does seem to be a general complicity between the EU's scientists and politicians that they will manage stocks for stability at whatever level they slump to next – some call it bumping along the bottom. Nevertheless, the best judgement has to be that if ICES advice had been taken for the North Sea's whitefish stocks – cod, haddock and plaice – in the past decade, as it was with the herring, the sea's fish populations would be in less of a desperate state.

The failure to act on what *is* known was just as bad after the war. When Michael Graham arrived back at Lowestoft in 1945, a considerable amount of thinking had already been done. In 1941 his mentor, Dr E. S. Russell, chaired a committee of scientists looking at ways to ensure that the fishing grounds of the North Sea would not be overfished after the war if the Allies were to win it. The committee made a number of astonishingly modern-sounding proposals. First was that there should be regulation of the number of days each vessel spent at sea. Another was that the tonnage of the fleet after the armistice should be set at 70 per cent of the fleet in 1938. A third was that there should be a minimum mesh size, but this should be larger than that agreed at an overfishing conference in March 1937.

Then a process of political dilution was applied to what the proposers had intended as a coherent package. The Admiralty and the Department of War Transport objected to the proposal to cut tonnage for military reasons. A war cabinet committee ruled in March 1945 that Russell's recommendations should be applied only to North Sea fisheries, not those in the North Atlantic and Barents Sea, as the evidence of overfishing there '*was not absolutely conclusive*' [my italics] and because negotiations would involve several countries, including the Soviet Union, so were unlikely to succeed. The recommendations on the North Sea, however, were accepted. The Foreign Office was instructed to convene an international conference on overfishing as soon as possible.

What happened at that conference in March 1946 is worth repeating because it so closely resembles what has gone on in Europe ever since. The United Kingdom pressed for the restriction of fleet tonnages. This was not accepted because it did not suit the post-war regeneration plans of other countries. The conference recommended instead that the fleets be kept at their present size, or that of 1938, whichever was greater. Even this measure was not acceptable to Denmark, Norway and Sweden, and was only accepted by Spain and Iceland on the understanding that it did not interfere with their plans for building new vessels. Ideas for a total allowable catch, close seasons and comprehensive fishing trusts, with exclusive rights over certain fisheries, all foundered. The only agreement from the conference was for a convention to regulate mesh sizes. It took until 1953 for all signatories to ratify the convention, after which a permanent commission was set up in London to run it, with a secretariat provided by the ministry. This commission was unable to resolve disputes, such as those that broke out between the United Kingdom, with its large, long-distance fleet, and Norway, Iceland and the Faroe Islands. The failure of these post-war arrangements led to the 'cod wars' of the 1960s and 1970s. Why did these arrangements fall apart? It was the same old story: fishermen vote – fish don't.

Thus was the great opportunity to manage Europe's fish stocks for future generations squandered. It will take the equivalent of a world war to bring the fish back again.

CHAPTER 8
AFTER THE GOLD RUSH

BONAVISTA, NEWFOUNDLAND. FALL. Nothing prepares you for the beauty. When I turned off the highway on to the Bonavista peninsula the road soon began to wind around the rocky coast through fishing settlements, known as outports, and the setting sun burst out of a stormy sky. The squat orange maples and yellow birch glowed brilliantly against stunted fir and spruce. Looking inland, vistas appeared towards the hills, over peat bogs and what in Scotland would be called black lochs. Seawards, looking along the rocky promontories and bays, the trees were few, but the clapperboard houses looked welcoming, and one inhabitant took great pains to show me the way. Wooden stages, or flakes, where cod was once dried, looked like relics of an earlier, more prosperous age. That, when you think of it, is exactly what they are – remnants of a 'gold rush' that lasted nearly 500 years.

Nothing quite prepares an English visitor for the distances in a province the size of England that was, until 1949, a separate country and has its own time zone, half an hour different from the Canadian mainland. The plane landed in St John's at 1.30pm in rain, low cloud and 112-kph (70-mph) wind, and I was drenched before I reached the hire car. It was 320 km (200 miles) to Bonavista by road. Having been warned not to drive after dark because of moose on the roads, I drove north up the main highway as fast as I dared. The radio was tuned to a station with a 1980s' playlist that included Tom Petty and Fleetwood Mac, varied occasionally with contemporary singers, such as Dido. This produced a sensation of drifting in and out of the past that I was not to shake off until I left Newfoundland. It was dark when I arrived in Bonavista and I was probably as grateful to see it as John Cabot was when he sighted the windswept cape from the sea in June 1497, exclaiming 'Buona vista!' (What a good sight).

My sense of being in a time warp was heightened further when I read the tourist information folder about Bonavista (compiled in the early 1990s) provided in Abbotts B&B on the road out of town towards the cape. Typed sheets of paper in plastic folders told me that the town is windy all year round, but does not have really cold weather or hot summer temperatures because it is at the end of a peninsula. The spring is short. Icebergs drift down the coast well into May, when the weather can be miserable. Only the brave swim in summer. Bakeapples, otherwise known as cloudberries, grow in the marsh. They got their local name from the French, who asked 'Baie qu'appelle?' ('What's that berry called?'), and are noted for their taste. So, too, are partridgeberries, which contain a natural preservative that means you can keep a pot of them in water all winter to bake in muffins. Youths amuse themselves by persecuting frogs in the marsh and driving around on quad bikes. The tourist information lists local issues and concerns: whether the Arctic ice will remain inshore into the fishing months, for if it does, it will lower incomes; fish prices; the availability of fish species; the increase in whale and seal numbers. Under the heading 'Daily life in the community' some thoughtful Abbott had recorded: 'The fishery is Bonavista's main industry. Cod drew the Europeans here in the 1490s and keeps us here in the 1990s. Without the fishery, Bonavista could not survive. IN COD WE TRUST.'

Nothing had been added to this passage since 1992, the year the Canadian government closed the cod fishery. Perhaps the author thought, as did the Canadian government of the day, that the crisis would be over in a couple of years. Perhaps no one in the Abbott household quite knew how to come to terms with the end of a period of dependence on the codfish that began nearly 500 years earlier with the arrival of John Cabot. Born in Venice as Giovanni Caboto, his real name was anglicised to stress the official nature of his mission. He had been commissioned by Henry VII to find a western passage to Asia. What he actually discovered in 1497 was a northern landfall on the American continent from which all Britain's subsequent claims in the New World derived, and cod in such quantities that, he claimed, all you had to do was lower a basket from the boat to catch them. At the time of Cabot, Canadian scientists estimate that the amount of spawning cod off the Canadian coast amounted to more than 4 million tons. By 2003 there was about 50,000 tons.

Remote places often have the warmest welcomes, and Bonavista prides itself on the friendliness of its people. I arrived in the dark and the rain on a Sunday evening to find my main contact had gone home. Next door was the town's café, a place that was open but wasn't quite finished yet. Its proprietor, Harvey Templeman, saw a tired, jetlagged figure with no Canadian money. He pushed across a cup of free coffee and busied himself with the phone directory until he had found the home number of the person I wanted from the mayor's office. Remote places can also be just plain strange. At the restaurant I found on Capeshore Road, nearly everything on the menu was fried. The waitress told me she was the common-law wife of a shrimp fisherman. Where was I from, she asked, not recognising the accent. The UK. She asked tentatively, 'In the UK does the water go the other way round the toilet?' This is a place that is nearer by several thousand miles to Dublin than to Vancouver. And nearly everyone's family was originally from the British Isles. I began to understand why mainland Canadians tell Newfie jokes.

I came to Newfoundland to try to understand for myself why the cod collapse happened, how it affected the place and what hope there was for the future. Some 44,000 people in fishing and processing are said to have been put out of work in 1992 – an enormous number. One wonders what exactly they were all doing. In the early 1990s there were 705 jobs in Bonavista directly provided by the fishery, in catching and processing. Next day I asked Betty Fitzgerald, the mayor, what happened to them. She told me in the local accent, which ranges between Ireland and Devon without setting foot on the Canadian mainland, that the town had lost 700 people since 1992, though not all were fishermen. Since then, inward migration had started again, with people moving up from Ontario and the United States. They had discovered the place through tourism, which hardly existed before.

The town has survived in the absence of cod by placing its trust in its heritage. Interest began with the 500th anniversary of Cabot's landing. A replica of his vessel, the *Matthew*, was finished in 1998 and now over-winters in a specially made boat-house in the harbour. Heritage trails were created along the Bonavista peninsula and in other regions of Newfoundland. Since then, an increasing flow of tourists has arrived to see John Cabot's statue and the lighthouse on the end of the cape, and mostly just to drive around. Tourism hasn't quite made up for fishing yet, but Betty is working on it. She has successfully sought 80 per cent heritage grants for historic buildings, of which Bonavista has over 1,000 – more than any other settlement in Atlantic Canada. The oldest is a house built in 1811 by Alexander Straithie, a tradesman from Renfrew, Scotland. Doing up old buildings provided employment for enough weeks for fishermen to be entitled to employment insurance for the rest. But Betty Fitzgerald still has 500 people seeking 200 jobs. She's getting a theatre started and trying to get a slate mine opened again. She wishes she could get an old sealing plant reopened. Bonavista is trying all the things that Cape Cod fishing communities did when they ran out of fish.

Bonavista has a lot of historic advantages compared with other parts of Newfoundland. After Cape Spear, the most easterly part of the continent just south of St John's, it is the most obvious historical site that any visitor might want to see. Only now the town is beginning to wake up to its potential, so I was surprised when Betty described the cod fishery as the backbone of the community. Her three sons still work in fishing or fish processing, although they have struggled over the past decade. There were still plenty of fish in the bay, she told me, so she couldn't understand why the inshore fishermen – who fish within the 20-km (12-mile) limit – couldn't use their hand-lines and traps to catch it. That was the strongly held local view, explained to me patiently by Larry Tremblett, an inshore fisherman with 23 years at sea, who now faces a fine of $500 (£208) if he catches a single cod.

The bizarre thing is that Bonavista Bay and Trinity Bay are full of cod. Sitting in his 10-metre (35-ft) boat *High Hopes* in Bonavista harbour, Larry explained: 'That's the way it has been round here for five years. When we put out lumpfish nets, or try to catch blackback [winter flounder] we get cod. We get cod of 50–60 lb [22–27 kg] sometimes. Nobody knows what the Department of Fisheries and Oceans is up to. They just don't want us to get at it. They think the inshore fishery is a pain and they are trying to freeze us out.'

There is bad blood between the inshore fishermen and the Department of Fisheries and Oceans (DFO), which is based in St John's and Ottawa. The DFO ignored warnings from the inshore fishermen in the 1980s that their catches were declining and that big draggers, as stern-trawlers are called, were overfishing the Grand Banks cod. The inshore fishermen were right, but the DFO scientists said it was too complicated to measure what the inshore fishermen did, so the fishermen's claims were never tested. Now the DFO says that the fish in Bonavista Bay and Trinity Bay are all that remain of what was once the largest cod population in the world. The northern cod once used to spread along the coast of Newfoundland and Labrador and 385 km (240 miles) or so out on the Grand Banks. (There are actually ten separately spawning stocks of cod off the Canadian coast, but the northern cod was always the largest.) Even though the DFO has almost certainly got it right this time, the inshore fishermen don't believe the DFO on principle. It's probably not such a bad principle, given what they've been through.

The history of the Newfoundland cod fishery looks different from the inshore fishermen's perspective. By the early 1990s Larry Tremblett and Doug Sweetland, for example, saw the population of cod begin to retreat south, until there was none off the coast of Labrador, then none off the north of Newfoundland. Next the inshore fishermen noticed that the fish were getting smaller. The Fishermen's Union, which represents the inshore fishermen, pointed this out to the DFO and called for a ban on dragging the spawning grounds. As Alan Christopher Finlayson has pointed out in his book *Fishing for Truth*, the DFO preferred to work with data from the big companies that ran the trawlers. 'They used to look at us as if we didn't know what we were talking about,' says Larry. 'We proved them wrong. If they had started listening to us in the 1980s, we wouldn't have had the problems we have today.'

Doug Sweetland is even-handed in dishing out blame. 'Everyone contributed to the decline of the cod. The main culprits were the draggers because of high grading.' High grading is where fishermen keep catching fish and throwing away the ones they don't want until they have achieved a full hold of premium-size fish. 'The plant in Catalina employed 1,500 people. They were getting so much fish in the mid-1980s that they said 26 inches [65 cm] is the smallest fish we want. You know where the rest was going – over the side. That destroyed a million pounds of fish to get 400,000 lb [180,000 kg].' Doug remembers how quickly the end came. 'In the winter of '92 there was good cod. Within three months there was nothing. The offshore fishery had shot itself down.' Ironically, it was the inshore fishermen, who caught least, who have suffered most. Of the big companies, FPI is back in business processing shrimp and crab, which it buys from boats that it does not own.

For those fishermen who used to catch cod close inshore, the future looks bleak. The state of the northern cod 12 years into the crisis is now worse than it was in 1992, thanks to another foul-up by the DFO and its political masters. Back in 1992, those scientists who recommended closure of the fishery predicted that the stock would recover to fishable proportions in two years. Working on that assumption, the Canadian cabinet decided to pay fishermen a $4 billion (£1.7 billion) 'package' – for social security, licence surrender and retraining – that eventually ran out in 1995. Still there was no recovery. The Treasury refused to subsidise the fishermen further, as the disaster was at least partly of their own making. Strangely, it made no such decision about the DFO.

As a result, fishermen in rural Newfoundland pressed for a small-scale reopening of the fishery. The government had run out of options. The fishery was reopened, contrary to the advice of many scientists led by Ransom Myers, by then no longer at the DFO. This so-called 'sentinel' fishery was meant to contribute towards scientists' knowledge of stocks by recording the age profile of the cod that was caught. Although it was restricted initially to traps and hand-lines, it was a commercial fishery. Owing to the number of fishermen participating, it proved hard to monitor catches or cheating. Catches were almost immediately unsustainable, according to Peter Shelton, a DFO scientist, who was responsible for monitoring the result. The sentinel fishery was closed down in spring 2003, after DFO scientists concluded that 'serious harm' had been done to stocks. The big difference was that this time the minister and the DFO accepted that there was no chance of things getting better in the near future. The best estimate of a recovery time was 15 years. Nobody actually knows if the cod will ever recover.

The second foul-up by the DFO has given Bonavista's inshore fishermen very few options. For although they might want to catch other fish, what they do catch when they shoot their nets or traps is cod. Larry was fishing for blackback in Bonavista Bay and worried himself sick when he inadvertently caught 1,350 kg (3,000 lb) of cod, until officialdom let him off. He now gets by fishing for crab for just a few months of the year. He makes enough in the 14-week spring season for crab to qualify for employment insurance (EI), around $600 (£250) a month, which the government pays for the rest of the year.

I found inshore fisherman Doug Sweetland at home after he had returned from moose hunting in the north. 'Got to scrape a living,' he grinned. Doug catches crab, lobster and lumpfish to qualify for his EI. Doug told me the extraordinary story of the time the remnant northern cod died of cold.

For some unexplained reason about 20,000 tons of cod, sizable fish of 3–5 kg (8–10 lb), shoaled during the winter of 2003 in Smith Sound, south of the Bonavista peninsula. The water was really cold, and perhaps because so many cod were gathered together, lots of them were pushed up from the bottom into the coldest water where ice was forming. The ice blocked the cod's gills and they weren't able to extract oxygen from the water, so they were smothered. Doug said: 'The scientists couldn't believe how cold the water was. Two million pounds [9,000 tons] of cod was just floating there, dead. You were allowed to pick it out.' It is not the only time this has happened: the first recorded occasion was seven years earlier. There is a plausible link with overfishing: cod and haddock are known to huddle together when they are depleted.

The ban on catching cod for the foreseeable future has sharpened the fishermen's resentment of seals and foreigners – usually, but not always, in that order. Doug calculates that the amount of cod caught by fishermen in the sentinel fishery amounted to 36,000 tons over the past six years. He calculates that hood and harp seals, of which there are 6 million or more on the Newfoundland and Labrador coasts, ate 50,000 tons of fish a year. 'The scientists said that didn't affect the stock. Only what we caught affected the stock. Do you believe that?'

Did seals really eat cod, a fish that lives on the bottom many fathoms deep, I asked Doug. In reply, he me told the story of a time he was out shooting terns – legal if for personal consumption in Newfoundland. 'We saw this seal come up in 100 fathoms [600 ft] of water with a fish in its mouth. The fish was alive. The seal let the fish go and we turned the boat back just to pick it up. It was a cod.'

The other all-too-real threat to the remaining cod comes from foreign fishermen, who continued to fish the 'nose and tail' of the Grand Banks outside Canadian waters – and possibly *in* Canadian waters at night. Reports by observers show 5,000 tons of cod, Greenland halibut and other moratorium species still being caught in the area managed by the North Atlantic Fisheries Organisation outside the Canadian exclusive economic zone. Betty Fitzgerald said she has been out on the Grand Banks since 1992 and seen the lights of foreign trawlers 'like a city at night'. It is a common reaction among fishermen everywhere to blame foreigners for overfishing. That does not mean they're wrong.

Some might say that the logical thing, if Bonavista can't catch fish, would be to make a virtue of necessity and declare the bay a marine reserve and make it a tourist attraction. This is not a popular option. Parks Canada, which runs the national parks, tried to impose just such a solution. The men in Ottawa miscalculated badly when fishermen found out, as the bill was going through its second reading in Parliament, that the area they had been told was excluded was not. 'What the guy was saying and what was happening were two different things.' You wouldn't have been able to control seals, said Doug, and when Parks Canada set up a national park on the coast in the 1940s, they burnt all the jetties and forced out the fishermen. Faced with implacable local opposition, the marine reserve did not get through. Doug has more time for a marine protected area, set up under the DFO, which a group of lobster fishermen have set up to protect the lobsters from overfishing in Trinity Bay. Doug wants to go back to catching cod. He thinks the stocks in Bonavista Bay would stand it if inshore fishermen like him had a quota of a ton a head.

Before I head off back to St John's, I drop in on Harvey Templeman again. He is outside, painting the shop in his overalls. We stand and look at the town: the harbour, the courthouse, the busy junction in front of us, where many of the trucks passing are those of semi-employed fishermen. 'You do wonder about these fishermen sometimes. They say they are hard up, but they're still driving around in big trucks.' Harvey knows what the outside world is like: he spent four years in New York City. He is a businessman, in a pretty laid-back way, and he can see the place changing. 'People are discovering that tourists are coming now. They are beginning to understand that it can be a business. There weren't any tourists before 1997.'

I tell him, because he asks what I've learnt, that I do find it an amazing fact that fishermen face fines of $500 (£208) for catching cod while there are loads of fish in the bay. His girlfriend says, 'It's not amazing. It's outrageous.'

Bonavista is dealing with the present. But, like the rest of rural Newfoundland, it hasn't given up on the past. It wants to catch cod, as it has always done.

*   *   *

Back in St John's the next day, I talk to Alastair O'Rielly, who is busy winding up the Fisheries Association of Newfoundland and Labrador – the organisation that represents fish processors and companies that operate the big dragger fleet – because the members are frustrated with the glut of capacity in the processing sector and their success in influencing government and want a new start. He explains that a gulf exists between fishermen in the north, where I have been, and scientists, that no one has yet bridged.

> The only remnants of the northern cod are in Trinity Bay and Bonavista Bay. The people in that area hold the view that the stock is healthy. No one else holds that view.
> This body of cod has been analysed to death. There is a spawning biomass of 50,000–60,000 tons where once there was 1.5 million. There's nothing left. We have 380 shrimp vessels fishing on the Grand Banks. We have no cod by-catch problem. We would love to have that problem because it would give us some indication of recovery. It doesn't exist. DFO do a survey with a shrimp trawl. It catches everything. They don't find cod out there either. It doesn't exist. There is no indication that recovery has begun or is even possible.

On the other hand, O'Rielly explained the extraordinary upside of the cod crash. Harvesters, as he calls fishermen, have enjoyed a windfall. Thanks to the absence of cod, which used to eat shellfish, there has been such a bloom of shellfish, snow crab and shrimp that the industry as a whole is making more money than it did in the late 1980s. Total production of fish has exceeded $1 billion (£417 million) in the past five years, while it was only $800 million (£333 million) in the late 1980s. A further advantage is that while the value has gone up, shellfish being worth more than cod, the tonnage has halved, thereby halving the shipping costs. The only people who haven't benefited from this bonanza are the inshore fishermen with limited access to crab and shrimp, which are caught mostly well out on the Grand Banks by the bigger boats. O'Rielly says: 'Nature abhors a vacuum. We're very fortunate. It could have filled up with something the world doesn't want.' There was a time, he says, when you would step on a snow crab if it landed on deck because there was no market for it. Again Newfoundland was lucky. All the marketing work in Japan had been done for Alaskan snow crab, which collapsed just as Newfoundland snow crab became available in large numbers.

The 'ecosystem shift' on the Grand Banks, which may or may not be reversible, has made Newfoundland the largest producer in the world of snow crab and the largest producer of coldwater shrimp. Catches of snow crab went up from 16,000 to 69,000 tons in seven years. The result for fishermen equipped to catch crab and shrimp was incomes at levels they would not have thought possible in 1992. There is only one problem. No one, as Bruce Atkinson, regional director of science, oceans and environment at the DFO in St John's admitted, knows what a sustainable catch of snow crab or shrimp is. What scientists do know is that shellfish populations are volatile. O'Rielly adds: 'If you want to worry in technicolor, you'd worry that something happens to the shellfish resource. People haven't given it that much thought.'

The DFO has tried to set catch quotas on a precautionary basis. In the case of crabs, only males that have reached terminal moult are taken. In the case of shrimp, about 12 per cent of the total biomass is taken each year, half of what is calculated to be sustainable. In theory, the catch could be doubled, but that would be taking more risk than the industry wants. O'Rielly says:

> If we'd had a wish list since the moratorium, it would have been for crab numbers to be up 400–500 per cent, shrimp up too, for the Canadian dollar to depreciate and for our competitors in Alaska to decrease production. And we would have liked to see consumer willingness to go on paying high prices for shellfish. That's what we asked for and that's what we got. We just don't know how long it will last. It makes you nervous. So much is riding on it, not only for the harvesters,

but also the communities where they live. There is not much else in the rural economy. We're it.

*   *   *

You would have thought that some people might actually be happy with the way things have turned out. To my surprise, O'Rielly, like everyone else, wants the cod back. This, I get the feeling, is not only because he thinks a more complicated ecosystem might be more stable. Catching cod just feels right in Newfoundland, whether you are an inshore fisherman or have a university degree. People feel happier with what they know and believe is natural. O'Rielly, like so many others, feels that the numbers of seals around is not natural. Each of the 6 million harp and hood seals, he says, eats around 1.3 tons of fish a year. 'Just as long as we understand this is a choice we are making. The consequence is that 12 years into this crisis it looks as if we have been replaced by seals and whales at the top of the food chain.'

The biggest problem Newfoundland has faced yet will come, paradoxically, if the cod does increase. If the cod gets more plentiful, it will eat the shrimp and the small, soft-shelled crab. As Bruce Atkinson of the DFO in St John's put it: 'You think things are bad now? You're going to have a time when *everything's* shut down.' Atkinson knows better than anyone the political pressures that situation can bring in a province where up to 8,000 people still work in fish processing and up to 14,000 in what the industry calls harvesting. He observes ruefully of 1995: 'There was no longer-term plan in the 1990s. At the first sign of a few fish, everyone wanted to get at them.'

That's the problem. So what makes him so sure that the same cycle is not going to happen all over again? Is there now a long-term plan? Two things have certainly changed since 1995: one is the government's undertaking to favour the precautionary approach, which says that the fish population must be preserved above the level at which recruitment is impaired; the other is the 1996 Endangered Species Act, which does not allow fishing of endangered stocks. But what size must the cod stock return to, assuming it ever returns, for fishing to be allowed again? How many fishermen should be allowed to fish it? Nothing appears to have been decided. You wonder if any lessons have really been learnt by the DFO, by politicians or by the fishermen themselves. Already there is pressure to reopen the cod fishery in the north of the Gulf of St Lawrence. Will the DFO get it wrong a third time? While individual scientists undoubtedly strive for objectivity, the DFO policy-making machine is well attuned to Newfoundland's style of politics

Although a visitor to Newfoundland, I realised that I had now begun to get the measure of its cosy, hand-out-ridden culture. Political success is equated with getting more money out of Ottawa and almost never with conserving fish. The time has come to say that Canada's system of employment insurance amounts to a massive subsidy to fishermen to stay where there are. The subsidy ensures that there is a fully equipped fleet ready to go fishing the moment there is even a small amount of fish to catch. Ransom Myers puts as his number one reason for the cod collapse the Canadian government's introduction of unemployment insurance (as it was then called) in the 1960s. 'The only equilibrium in a subsidised system is zero fish,' he says. 'The system is set up to fail – necessarily.'

Myers studied the fishery from records made in the 17th and 18th centuries. He found that when catches were good, people moved into Newfoundland; when catches were bad, people moved out. Four centuries ago, each boat could catch and process 10 tons of fish, salting and drying it, which was a very labour-intensive process. Four centuries on, many boats were catching exactly the same amount. The efficiency had not changed at all because of subsidy. With employment insurance, it is not inconceivable that a couple can make $60,000 (£25,000) a year, spending 14 weeks catching or processing crab or other fish and the rest of the time improving their lives. For fishermen, all business expenditure is allowable against tax. The time they don't spend fishing can be spent fixing their home, shooting moose for food, swapping labour with their friends, or chopping wood. People enjoy hunting and gathering, provided their basic income is assured. An enviable life in many ways – but why, you may ask, should the taxpayer pay for it?

'It's totally crazy,' says Myers.

Too many people in the fishery create tremendous political pressure for rules to be bent and quotas to be increased. As a former DFO man, Myers says you won't hear a peep of criticism of this state benefit system coming from the DFO's big, modern building in St John's. 'For an academic in Newfoundland to say anything against employment insurance is totally unthinkable. It's not even discussed.'

Why do people, all over the world, think that there is something so valuable in coastal settlements that mean they must be subsidised? Why are people – certainly European and Canadian people – willing to subsidise fishing to an extent they might not contemplate with other industries, such as coal or even farming? Myers thinks it is because the hunter-gathering lifestyle appeals to something in all of us. That may be true, but the justification for it, post-cod crash, is definitely past its sell-by date.

Myers is not entirely right about the DFO failing to grasp the nettle of subsidies. A rare example of it acknowledging the pervasive influence of subsidy comes in a paper by Jake Rice, Peter Shelton and three other scientists, looking back at the cod crash and forward to the future. It sounds increasingly like Myers himself. In it the scientists point out that the auditor general of Canada investigated the $4 billion (£1.7 billion) Atlantic Fisheries Adjustment Package, the social support scheme that existed until 1995. The package paid minimum income support of $400 (£166) a month to fishers and plant workers who had lost their jobs. The package paid double this to workers who agreed to retrain and learn skills, such as computing, which were readily transferable back into fishing. The auditor general found that most of those who had been retrained said they would return to fishing as their primary job, just as soon as it was possible to do so. Profits from the new crab and shrimp fisheries tended to be reinvested in technologically sophisticated vessels equipped to take part in a range of fisheries, including cod, if there were any to catch. The result was that after spending $4 billion (£1.7 billion) to 'adjust' the Atlantic cod fishery, the effective fishing capacity was 160 per cent of what it had been in the early 1990s.

When fishing resumed, Rice and his fellow scientists continue, the imperative resulting from so many fishermen needing 14 weeks' income so that they could claim employment insurance made the participation of the maximum number of people, rather than profitability, the governing factor in the allocating quota. A large number of very low quotas were therefore doled out, which proved hard to monitor and enforce. This means a lot of cheating went on.

The men at the DFO do seem at last to have recognised that the costs of over-optimism are greater than the costs of being unduly pessimistic. They say the prospects of a cod recovery in the next few years are very unlikely. Even small fisheries could annihilate any gains and prevent the stocks from returning to their historic biomass and yields. Although the wily DFO men could just be distancing themselves from blame when Canada gets it wrong all over again, I think their summing-up has the ring of truth. Rice and his co-writers conclude: 'At least in Canada, fishing is a culture, not just an economic activity. People will come back to the fishery when it reopens, in large numbers and with new skills and high expectations.' They say that unless excess capacity (which means fishermen and fishing vessels) is *permanently* removed from the fishery early in the recovery, the expenditure on science and management needed to ensure the stock recovers will turn out to have been pointless.

No one should underestimate the political difficulties of stopping fishermen from fishing in a province where so many of them vote. Yet somehow it must be done, or the cycle will go on repeating itself. The cod will recover – just, maybe – and will be fished out all over again. Until someone has the courage to strike out in a new direction, Newfoundland will just be a desperately sad place, making the same mistakes over and over again without ever appearing to learn from them.

SUBSIDIES

A subsidy is a sum of money dished out by the government, which allows a commercial venture to go on doing something beyond the point at which it would otherwise have gone bust or been forced to do something else. Despite the increasingly alarming state of the world's fish stocks, there are plenty of countries still doling out subsidies to fishermen that will ultimately make things worse for fish – and therefore for fishermen.

Initially, the taxpayer might benefit from plentiful supplies of cheap fish, but he or she will end up paying twice – once to pay fishermen to fish and again when the price of fish rises because stocks have been overfished. The unfortunate taxpayer may even have to pay a third time, as in Canada, when the fishermen have to be paid *not* to fish because stocks have disappeared. Canada, however, is by no means the worst in the subsidies racket.

Globally the amount of subsidy in fishing is very large – as much as a staggering US$50 billion (£27.8 billion) a year, according to one estimate by the UN Food and Agriculture Organisation. More conservative estimates put it at up to US$20 billion (£11.1 billion). The difference exists because what qualifies as subsidy is a matter of dispute. Industrialised countries alone gave US$6.7 billion (£3.7 billion) of subsidies to their fishermen in 1996 according to the Organisation for Economic Cooperation and Development (OECD). This fell to US$5.9 billion (£3.3 billion) in 1999, as more countries realised that subsidies might be the root of the overfishing problem.

Top of the subsidies league is Japan, which handed out US$2.5 billion (£1.4 billion) to fishermen in one way and another in 1999. It is followed by the European Union, which hands out US$1.16 billion (£644 million), excluding Belgium and the Netherlands because they were slow with their figures that year. In third place is the United States, which doles out

US$1.11 billion (£617 million). Individual EU members – Spain, France, Ireland and Italy – all make impressive showings in their own right, doling out US$345 million (£192 million), US$139 million (£77 million), US$104 million (£58 million) and US$92 million (£51 million) respectively (1997 figures). Norway, another big player, despite its tiny population, hands out $163 million (£91 million), which works out much more per head.

Subsidies amount to a staggering proportion of the actual value of landings. In the USA it was over 30 per cent in 1999, in Japan 24 per cent and in the EU 17 per cent. Two decades ago the Soviet Union handed out free boats and free fuel to its fishermen – a big but unquantified subsidy. There remains a total lack of transparency everywhere about fishing subsidies and who gets them. The only public audit of them ever conducted, by the European Court of Auditors, found numerous payments to people who did not qualify for subsidies, including at least one to the owners of a vessel that had sunk.

What qualified as subsidies in the OECD's study included direct payments, such as grants for building new vessels and for making old ones safe, price supports for fish, grants for companies to set up joint ventures with other countries, tax exemptions and interest rebates for the purchase of vessels, and general services, such as state expenditure on fisheries science, research, management and law enforcement. Buying access for fishermen to other countries' waters, from the north of Norway via Africa to the Falkland Islands, accounted for a sizeable US$430 million (£239 million) of the European Union's US$1.1 billion (£611 million) expenditure on subsidies.

Some countries now insist on recovering a significant percentage of the government's costs from the fishing industry. New Zealand, Iceland and Australia, for example, recover 50 per cent, 37 per cent and 24 per cent respectively of the cost to the public of fisheries research, management and enforcement. These countries, arguably, run some of the best-managed fisheries, so there seems to be a connection between no subsidy and good management.

The converse is also true. Subsidies create over-capacity in the industry. The global fishing fleet is estimated to be two and a half times greater than needed to catch what the ocean can sustainably produce. A big fleet creates pressure on politicians to use subsidies to buy licences for fishermen to fish elsewhere if their own waters are overfished. In Europe it is those countries that subsidise their fishermen the most that lobby for this over-capacity to be exported to the rest of the world. (Spain, for example, spends US$196 million [£109 million] on decommissioning vessels, which allows the modernisation of its fleet.) Spanish and Portuguese fishermen, who used to be subsidised to fish in Moroccan waters, were *compensated* by the European taxpayer to the tune of another €197 million (£131 million) when they were chucked out after Morocco refused to renew one of the EU's controversial fisheries agreements.

Some people believe that there are good and bad subsidies. Half of all fishing subsidies received by Spain from the EU are environmentally damaging, according to one WWF report. The authors of this report strained their credibility by finding that 36 per cent of subsidies were actually good for the environment, and 15 per cent were neutral. Generally, I believe that if you look hard enough, you find a subsidy nearly always harms the environment or disadvantages someone somewhere else, usually by distorting prices in favour of fishermen in the north, which penalises the south. Even so-called neutral subsidies, such as those paid to improve the safety of boats, mean that fishermen will stay out longer in bad weather and catch more fish.

So what lies at the root of a democratic politician's impulse to dish out subsidies? First, a disgraceful need to buy votes with other people's money, often dressed up as the redistribution of wealth. Second is the misguided belief that subsidising fishing is somehow *investing* in the industry. In fact, in a hunter-gathering economy, you invest only by leaving the resource alone. The way to tackle subsidies in well-governed countries is to create transparency, a free press and proper scrutiny by public auditors. That way people get to realise that farmers and fishermen are walking off with their money for no good reason.

One of the commendable aspirations of the round of free trade talks that began in Doha in 2001 is to reduce fishing subsidies of all kinds. Despite the addiction of most environmental bodies to obstructing the World Trade Organisation's processes and carping about its failures, most recognise that this would be an unreservedly good thing.

CHAPTER 9

LAW AND THE COMMONS

ATLANTIC DAWN is an evocative name for a vessel. To a European there is something curious about it, too. If you live in the British Isles or on the mainland of Europe, the sun comes up over the land, and east is never over the Atlantic, unless you are at sea or on an island. Kevin McHugh, the owner of the world's largest super-trawler, began his career as a fisherman-entrepreneur on Achill Island off the west coast of Ireland, so quite conceivably the dawn did come up over the Atlantic for him. I remember reading in the *Irish Times* that when his beautiful new boat was built, Mr McHugh planned to take the first opportunity he could to anchor off Achill island, so his mother, Nora, could take pride in his new venture. There was great charm in the portrait of this unassuming, private, softly spoken man who had become Ireland's most successful fisherman. You did have to remind yourself that the object of his pride was the greatest fish killing machine the world has ever seen, a vessel of 14,000 tons with a crew of 100 to catch, pack and freeze the fish that were stored in its cavernous hold.

Kevin McHugh made a career by spotting opportunities. He bought his first boat, the 65-foot *Wavecrest*, in 1968 at the age of 21 and was the proud owner, eight years later, of the *Albacore*, built partly to his design, at a cost of £1.2 million. The closure of the Irish herring fishery led to Killybegs on the Donegal coast where he concentrated on another pelagic fish, the mackerel. Then, like many other Irish skippers, he moved to bigger and bigger boats until in the late 1980s he placed an order many thought would break him, for a £12 million vessel that could steam even further to meet the demand for fish. The *Veronica*, named after his wife, was a super-efficient purse-seiner designed to catch fish inside and outside European waters as fisheries near to home began to drift into crisis. On the way, however, he suffered a setback. The original *Veronica* burnt out in a fire in Harland and Wolff's shipyard in Belfast while being repaired. It was eventually replaced, in 1995, with a larger *Veronica*, 104 metres (345 ft) long and, at 5,206 tons, some 1,172 tons greater than its predecessor.

With the *Veronica* fishing off Mauritania, McHugh appears to have seen further opportunities in the plentiful stocks of sardinella and other pelagic fish off West Africa, already pursued by a Dutch fleet of vast super-seiners. The vessel he commissioned to compete with them, the *Atlantic Dawn*, was built in Norway, making use of the subsidies available there for shipbuilding. The giant 145-metre (480-ft) purse-seiner and pelagic trawler was built with £4 million in subsidies from the Norwegian government out of a total cost of £50 million. It was designed to catch, process and freeze up to 400 tons of fish subsidies available there for shipbuilding. The giant 145-metre (480-ft) purse-seiner and pelagic trawler was built with £4 million in subsidies from the Norwegian government out of a total cost of £50 million. It was designed to catch, process and freeze up to 400 tons of fish a day and to accommodate 7,000 tons in its hold.

McHugh told the *Irish Times* he was unhappy with the perception that such powerful ships posed a threat to the world's fish stocks.

Where we will be working, off Mauritania, there are strict controls and very strict arrangements, whereby we must take a certain number of Mauritanian fisherman on board, and an observer. They like to see us coming, because we are paying for our investment and there is a direct return to the state. Those waters down there are alive with fish – like Bullsmouth – but the air and sea temperatures make it very difficult to land quality. That's what we aim to do, and to do so in co-operation with the Mauritanians, so both of us can benefit.

One of the reasons for the vessel's size, he explained, was about having the horsepower to drive an onboard factory with sufficient freezing capacity to ensure the quality of the catch. His company would be selling the fish on to the African market, a far better thing than the Russian vessels that were catching the same fish for fishmeal. Mr McHugh's financial package was put together by a syndicate of Irish banks.

The only problem came when the *Atlantic Dawn*, all sleek and lovely, was about to sail triumphantly for Dublin, when it was revealed that it did not have a licence to fish in EU waters – or any other waters, come to that. As the Irish Green party MEP, Patricia McKenna, put it at the time: 'Mr McHugh has essentially said to the Irish Government: "I have this boat, please find me some place to go fishing."' Mr McHugh was not alone in making that request. The syndicate of banks that had lent him the money was asking too.

The *Atlantic Dawn* was a big project for a small country. People in government circles say it had developed such momentum it could not be allowed to fail. The Irish government did its best to find it somewhere to fish. It applied at the end of 2000 to increase its pelagic fleet on the grounds that there were as-yet underexploited fishing opportunities in West African waters. The European Commission pointed out that Ireland's domestic pelagic fleet was already 40 per cent larger than it should have been under a Europe-wide fleet limitition to which member states were supposed to adhere. This, the Commission said, was illegal under EU law. No increase could be given unless the Irish authorities acted to clear the excessive fishing power of the pelagic sector. While all this argy-bargy was happening, Frank Fahey, minister of the marine, gave the *Atlantic Dawn* a place on the Irish merchant register – though EU rules say that fishing vessels are required to be registered on member states' fishing register – and granted it several temporary fishing licences. So the largest fishing vessel in Ireland went fishing as a merchant vessel for a year and a half. This, as the European Commision pointed out to the Irish government who had sanctioned all of this, was illegal and, in November 2001, the Commission opened infringement proceedings against Ireland. As an Irish television investigation subsequently revealed, even the *taoiseach*, Bertie Ahern, intervened in an attempt to break the impasse. He wrote to the head of the European Commission, giving the impression that if the *Atlantic Dawn* company were to go bust, it would have serious consequences for peace on the troubled border with Northern Ireland. Representations were even made to the EU fisheries commissioner, Franz Fischler, by Ireland's EC commissioner, David Byrne, responsible for consumer affairs who, although accepting that he had raised the matter with Fischler, stated that it was in the normal course of business and he had done nothing wrong in doing so. The result of all this was that to the dismay of a number of EU officials who pointed out that national fleets were meant to be getting smaller rather than larger, a deal was done that got the *Atlantic Dawn* onto Ireland's register of fishing vessels. The EU once again demonstrated how it appeared to be more interested in the regional politics of its member states than the sustainability of its fish stocks.

Ireland, however, did rectify what the Commission called the 'overhang' in its fleet capacity by removing the *Veronica* from its fleet and de-rating the power of another two fishing vessels. Though the *Veronica*, prior to the *Atlantic Dawn* the largest fishing vessel in the Irish fleet, was not actually going to stop fishing, the *Atlantic Dawn* was allowed on to the EU register. The reason given by a senior commission official for allowing Ireland more scope to fish was that scientific advice showed there was plenty of pelagic fish to catch in Mauritanian waters – the Dutch super-seiners hadn't been taking up their quota. The *Veronica*, totally legally, now fishes in Mauritanian waters under a Panamanian flag, on a licence owned by a Mauritanian company.

Dr Euan Dunn, a fisheries expert working for the Royal Society for the Protection of Birds, believes that the deal that allowed the *Veronica* to carry on fishing under a flag of convenience was a disgrace. Under a UN Food and Agriculture Organisation (FAO) action plan and under its own community action plan, the European Union was already committed to discouraging EU vessels from registering to flags of convenience that fail to fulfil their flag state responsibilities. Panama used to be one of the worst offenders against international fisheries agreements among the flag of convenience states, and although it has taken steps to remove some illegal tuna boats from its register, it must be seen as on probation at best.

This is not the only time that the European Commission has allowed the real killing power of the European fleet to expand. In another example, the commission accepted the size of the Dutch pelagic fleet at twice the size it was supposed to be under a previous EU decommissioning round – unleashing a total of 94,000 tons of technologically advanced fishing capacity, representing several highly equipped super-trawlers and purse-seiners, upon the waters of the North Atlantic and the rest of the world. It might, understandably, have been trying to keep control over a fleet fishing mostly in Mauritanian waters, which would otherwise simply defect to flags of convenience. But as far as reducing and containing fishing capacity goes, as Michael Graham and E. S. Russell first tried to persuade Europe to do 60 years ago, it has achieved very little.

What conclusions are to be drawn from the intriguing episode of the *Atlantic Dawn* and the *Veronica*? First, that fishing registers in developed countries of the world, now all supposed to be limited and contained under an FAO action plan, in fact leak like a sieve. (Old habits die hard. Not long ago, there were actually subsidies for moving off the EU register to a flag of convenience.) Second, that fishing capacity continues to grow everywhere. The Atlantic Dawn Company, which owns the vessel of that name and the *Veronica*, now advertises anchovies caught in European waters by the *Atlantic Dawn*, as well as a variety of pelagic fish caught off West Africa. If proof were needed that the *Atlantic Dawn* was contributing entirely legally to the global destruction of fish stocks, it is to be found on the company's website. One of the species on offer, is the unregulated and spectacularly overfished blue whiting.

It had been assumed until very recently by the FAO that fishing vessels switching to 'open' registries, where you can buy registration in return for a nominal fee, were old vessels, fully depreciated and nearing the end of their productive lives. Recently, however, someone in the FAO woke up and actually asked Lloyd's Maritime Information Services about this, and discovered it was not the case. An increasing number of young and newly constructed vessels have moved to 'open' registries. Ironically, the European Commission, which pushed the *Veronica* on to a flag of convenience, has pointed out that fishing by flag of convenience vessels represents a threat to the survival of fisheries worldwide.

So what exactly is the problem about flags of convenience? Ship owners have used them for years to avoid taxes and to cut labour costs, safety requirements and training for crews. Corner-cutting on health and safety hasn't gone down too well, particularly after some messy tanker accidents, but Panama and Liberia now have vessel inspections as stringent as anyone else's. By and large, the world has got used to a low-cost, lowest-common-denominator market in shipping. Indeed, some European countries, such as Germany and the Netherlands, even operate second registries with tax-free status on the grounds that if you can't beat 'em, why not join 'em. The problem with flags of convenience is that they often allow fishing vessels to avoid conservation agreements for fish. Under the UN Convention on the Law of the Sea, ships sailing the high seas are subject only to the jurisdiction of the flag state. Only within a country's 320-km (200-mile) limit or in other exceptional circumstances may a vessel of another country board or otherwise inspect a ship on the high seas. The flag state must therefore enforce all conservation agreements, including fisheries conservation and management agreements by regional fisheries organisations. Some do, some don't. Arguably, there isn't a problem as long as the flag state observes the fishing agreement. At least, that is what's claimed by the Spanish tuna fleet, which sails entirely under flags of convenience. Julio Moròn, assistant director of the Spanish purse-seiners organisation, told me that the Spanish fleet was under flags of convenience purely for the tax advantages. People who have fished on Spanish purse-seiners in the Indian Ocean say otherwise. They say a flag of convenience enables skippers to follow the fish unencumbered by the bureaucracy and quota that an agreement with an EU country involves. A skipper can make an arrangement with a particular African country on the spur of the moment and negotiate his own private fishing arrangements by fax.

I heard of one skipper in charge of a Spanish-owned vessel who thought he had bought a licence to fish in Somalian waters. Somalia, you will recall, is the only country in the world without a government at present, so he did this by agreement with one of three warlords who have effective control. When the vessel began fishing for tuna the crew were horrified to find themselves being overtaken by British mercenaries in a fast Russian speedboat. The mercenaries pulled alongside and told the tuna boat it would have to pay €200,000 (£143,000) into an account in Chelsea or it would be towed into Mogadishu and impounded, which meant effectively stripped. Money was transferred very fast and the British motored away. Apparently the purse-seiner had contacted the wrong warlord for the area he was fishing in.

There is always somebody prepared to accept money in return for allowing a vessel freedom to fish. A change of registration can be carried out by fax, over a satellite phone, while at sea. At the bottom of the cascade of money and responsibility there are many flag states that either do not enforce the treaties they have ratified, or never signed up to them. A basic principle of international law is that if a country does not adhere to a treaty, it is not bound by its provisions. A vessel flying the flag of Belize, for example, may fish perfectly legally for tuna in the Atlantic without paying any attention to conservation measures laid down by the International Commission for the Conservation of Atlantic Tunas (ICCAT). This amounts to a gaping loophole in the law, which appeals to those who want to fish on the high seas with impunity, and to those who wish to fish illegally in other countries' exclusive economic zones (EEZs) under the appearance of fishing on the high seas. The popularity of particular flags changes as international pressure is brought to bear on certain countries to remove miscreants from their registers – as ICCAT has made certain central American states do through trade sanctions. But there remains a handful of flags that appeal to companies that gear up to fish illegally for high-value species because they calculate that the risks of being caught are outweighed by the profits if they get away with it.

Pirate fishing of this kind can be every bit as cruel as the 2003 film fantasy *Pirates of the Caribbean*. Hélène Bours of Greenpeace International was sailing off the coast of Sierra Leone in the MV *Greenpeace* on 11 September 2001. She had already seen several vessels in a very dilapidated state, apparently lacking the most basic safety equipment and with the crew living in deplorable conditions. That morning the MV *Greenpeace* picked up a mayday call from a ship in distress: the *Estemar 5*, a Korean-owned vessel, registered to the Tikonko Fishing Company of Sierra Leone. The Greenpeace vessel headed immediately to its aid. Hélène Bours recalls:

> Shortly before we reached the location given in the distress call, about 2½ hours later, we saw a large slick of oil in the water, as well as some mooring ropes, plastic boxes and many other types of debris – presumably from the ship in distress.
>
> We set up look-out groups on the bridge and in the crow's nest and steamed about looking for survivors. A few hours later, a small boat with an officer of Sierra Leone's navy informed us that nine crew members of the fishing boat were missing. Only the captain and one deckhand had so far been found alive. We launched two of our inflatable boats to expand the search and to have a better chance of seeing any survivors. After another hour, without having seen any survivors or bodies, we terminated the search operation. Astoundingly, most of the other fishing vessels in the area had not even stopped fishing to help.

The *Estemar 5* was a serial offender. It had been sighted five times in a year in prohibited areas of the Sierra Leone 320-km (200-mile) limit during flights carried out by the Surveillance Operations Coordination Unit in The Gambia.

Very slowly the grey areas of the law of the sea are turning to black and white. The 1995 UN fish stocks agreement, which only came into force in 2003, requires flag states actively to comply with fishing agreements on the high seas, rather than merely not to undermine them. It also confers considerable powers on regional fisheries organisations to inspect and carry out surveillance of vessels fishing in their area, which includes allowing them to insist that satellite transponders be fitted. A voluntary plan of action agreed by the international community in 2001 provides support for flag states by tightening up scrutiny of the vessels on their registers and of port states – countries where the vessels dock – that collect information on fishing and deny access to flag-of-convenience vessels. As yet, however, the law of the sea still lacks the power to allow anyone but the flag state to impose measures on vessels on the high seas. Nor is there a way of forcing countries with open registers to observe the requirement, repeatedly stated in international law, that there should be a 'genuine link' between fishing company and flag state. That would mean that offenders could be easily prosecuted.

No edges are sharp when it comes to the law of the sea because it is always in someone's interest, even in the most apparently well-governed nations, for them to be fuzzy. The consensus treaties that make up the law of the sea are full of loopholes because they were negotiated that way. It is an ancient quandary, so a little history might make things simpler to understand.

The Romans believed that what no man controls, no man can own. Justinian, writing in the sixth century AD, said that the air, flowing water, the sea and the sea shore were common to all. In the 15th and 16th centuries Europe's crowned heads made elaborate and unenforceable claims over vast areas of sea and land. Venice claimed the Adriatic, England the North Sea, the Channel and part of the Atlantic, Spain the whole Pacific, Portugal the Indian Ocean and much of the Atlantic. Papal bulls, which Spain and Portugal claimed as their authority for laying claim to huge tracts of ocean, were unpopular and attracted opposition from nations who asserted the right of free passage. The case against these papal bulls was stated by Hugo Grotius, a Dutch jurist, in an essay entitled *Mare Liberum* ('The Free Sea', 1609). He insisted, on the basis of classical legal precedent, that the sea was *res communes* (common to all nations). John Selden, an English lawyer, historian and antiquary, riposted on behalf of the English crown with a critique called *Mare Clausum* ('The Closed Sea', 1635), which asserted the state's right to claim sovereignty over the seas adjacent to its territory. The practice of the Dutch herring fleet of fishing within sight of the English shores is understood to have been one of his reasons for doing so. In fact, by 1625 Grotius was already conceding that the common sea principle might not apply to the adjacent sea. The distance over which the state was held to exert effective control became established in the 18th century as the distance a shore battery could fire, then 5 km (3 miles). As gunnery improved, there was no attempt to extend this range, although some countries asserted sovereignty over 20 km (12 miles) to police matters such as smuggling. The 5-km (3-mile) limit lasted into the 20th century.

The tension between the common sea and the closed sea remains, however, long after the 1982 UN Convention on the Law of the Sea laid out the legal basis for each country to establish a 320-km (200-mile) EEZ. (If the distance between nations is less than this, the EEZ stretches to the median line between the two, i.e. down the middle of the North Sea.) The 1995 UN Straddling Stocks Agreement, which allows members of regional fisheries organisations to inspect vessels, has extended enforcement of a kind to the high seas. It is now theoretically possible, through satellite tracking and overflights by fisheries inspectors, to see and apprehend vessels that are fishing in the wrong place – much more like the system of de facto power that coastal states exercise in their own waters identified by Seldon – though not to enforce them on nations that are not party to the agreement. But although some fisheries, for squid, tuna and toothfish, are now global, regional fisheries bodies remain exactly that – regional – and lack power. In large areas of the ocean beyond the 320-km (200-mile) limit, such as on the Madagascan Ridge, there are rich fisheries with no regional organisations at all, and anything goes.

What happens in areas of ocean such as this is the purest form of what has become known as the 'tragedy of the commons', after an essay published in *Science* in 1968 by Garrett Hardin, an ecologist at the University of California, Santa Barbara. Hardin claimed modestly that he had merely renamed a phenomenon first described by the 19th-century British political economist William Forster Lloyd, who mused over the devastation of some common pastures in England compared with enclosed land.

The tragedy of the commons, as defined by Hardin, works like this. A pasture is open to all. Assuming each human exploiter of the common is guided by self-interest, he will try to keep as many cattle as possible on the common. This arrangement may well work for years because war, poaching or disease will keep the numbers of men and beasts below the carrying capacity of the land. At the moment when man and environment come into equilibrium, tragedy begins. Each herdsmen will ask himself what he is to gain by adding one cow to the common, compared with the loss he will make by overloading the pasture. The gain will always be greater to him, and the loss, because it is shared, less. And so it goes on, with every herdsman buying himself another cow until the common property is ruined and the cows starve.

Those who nod sagely and quote the 'tragedy of the commons' in relation to environmental problems from pollution of the atmosphere to poaching of national parks tend to forget that Garrett Hardin revised his conclusions many times over 30 years. He recognised, most importantly, that anarchy did not prevail on the common pastures of medieval England in the way he had described. The commoners – usually a limited number of people with defined rights in law – organised themselves to ensure it did not. The pastures were protected from ruin by the tradition of 'stinting', which limited each herdsman to a fixed number of animals. 'A managed commons, though it may have other defects, is not automatically subject to the tragic fate of the unmanaged commons,' wrote Hardin, though he was still clearly unhappy with commoning arrangements. As with all forms of socialism, of which he regarded commoning as an early kind, Hardin said the flaw in the system lay in the quality of the management. The problem was always how to prevent the managers from furthering their own interests. *Quis custodiet ipsos custodies?* ('Who guards the guardians?')

Hardin observed, crucially, that a successful managed common depended on limiting the numbers of commoners, limiting access and having penalties that deterred. The important thing about a successful common in medieval England is that there was not free access. The number of commoners on an English common were often as few as a dozen. Hardin points out that even when herdsmen understood the consequences of their actions, they were generally powerless to prevent damage without there being an adequately coercive way of controlling the actions of each individual.

If any group were able to make commons work on the basis of Marx's principle of 'from each according to his abilities, to each according to his needs', Hardin said, it should be an earnest religious community, such as the Hutterites in the western United States. Hardin observed that when the number of people in the community approached 150, individual Hutterites began to under-contribute according to their abilities and over-demand for their needs. He concluded 'numbers were the nemesis'.

At sea, where a common exists in most waters governed by nations or regional fisheries organisations, stinting remains the most favoured form of management. Where stinting doesn't work, which is in more places than not, this is generally because there are too many people involved for there to be trust between fishermen, or the penalties are not sufficient to deter them from following their financial interests. If there is to be a way of successfully managing a common sea, there needs to be a strong chance of being caught and tough penalties when an offender is convicted.

None of Hardin's requirements for a successfully managed common are fulfilled by high-seas fishery regimes. Nor very often do they work within 320-km (200-mile) limits either. The declaration of such limits still gives each nation the problem of allocating rights and setting adequate penalties. Frequently the share is resented if it is thought that some individuals have begun to cheat. When the well-meaning commoners, the bedrock of a working community agreement, perceive that others are cheating, they begin to cheat too. One of Hardin's lessons is that a common can evolve from a managed one to a tragic one. 'With an unmanaged common,' he wrote, 'ruin is inevitable.' A ruin with which the world is particularly familiar even contains the word 'common' in its title – the European Union's Common Fisheries Policy.

In later years Hardin confronted the fisheries problem directly. He wrote: 'If each government allowed ownership of fish within a given area so that an owner could sue those who encroach upon his fish, owners would have an incentive to refrain from overfishing.' Hardin was an early and influential advocate of granting property rights to fishermen to persuade them to conserve the resource and to police each other.

There can be little doubt that one of the most perfect examples of an unmanaged common on the planet is the Southern Ocean around Antarctica – the vast majority of it, at least. There ruin looms for the Patagonian toothfish, a large, bottom-living predatory species that can grow up to 2 metres (6½ ft) long and live for 50 years. The reason is not that there are no rules, but that the rules are so difficult to enforce in a vast, remote ocean with the highest winds and largest waves on Earth.

As with other relatively inaccessible deep-water stocks, fishermen only turned to the toothfish after other valuable southern hemisphere fish, such as the austral hake and golden kingclip, were eclipsed. Once consumer resistance was overcome by renaming it Chilean sea bass – perfect washed down with a glass of Chilean chardonnay – Patagonian toothfish sold well in the white-tablecloth restaurants of the USA and Japan. Its flaky white flesh had a pleasant flavour and stood up to being overcooked as well as cod. The price of toothfish has risen on the back of limited supply, since fishing for toothfish peaked in the mid-1990s, wholly justifying its description as 'white gold'.

The rewards for catching Patagonian toothfish, or Antarctic

toothfish – a less numerous species limited to the Ross Sea and shelves around Antarctica – are very high. The risks of being caught are very low. On the back of the current price, it is possible for a toothfish poacher to catch enough to pay for his boat, pay his crew and make US$500,000 (£278,000) profit in a single trip, according to David Agnew of Imperial College, London, who manages the toothfish stocks around South Georgia on behalf of the South Georgia government. No wonder illegal operators are willing to risk their vessels being impounded. If that happens, they do not hang around to get them back.

With at least 33 large vessels fishing illegally in the southern Indian Ocean each year, and only one or two arrests a year, 'the odds are better than people smuggling or drug running', according to David Carter, chief executive officer of Austral Fisheries, based in Perth, Western Australia. Carter's company runs two boats trawling legally for toothfish. In Australian waters poachers met a new phenomenon: legal fishermen with ownership rights to their fish who were prepared to campaign to protect their assets.

With the toothfish two major ecological problems have intertwined, which, ironically, have helped the legal fishermen to make a case for tougher action by governments. First, of course, the toothfish itself is chronically overfished, with illegal fishing taking more than half annual catches, and 80 per cent by some estimates. Second, the long-lining methods used by the fishermen who fish for it illegally in the Southern Ocean have been wiping out albatrosses; in fact, 17 of 24 species are now under threat of extinction. Australia has got the measure of bird by-catch in its waters – 320 km (200 miles) around its remote Southern Ocean territory of Heard Island – by banning long-lining in favour of trawling. The British dependency of South Georgia has cut bird deaths to perhaps 20 a year by requiring long lines to be released at night or under multi-coloured deterrent streamers. In the rest of the ocean it is up to the fishermen themselves what measures to take and, being illegal, many choose to take none.

The plundering of the seas around Antarctica is not new. British and Norwegian whalers killed 30,000 blue whales in their most successful year, 1929–30. Now these gentle giants, the largest creatures on Earth, are reduced to a total global population of 1,500 at most. The Soviet fishing fleet, once the largest in the world, used to fish in the Southern Ocean for a variety of stocks, including icefish and the marbled rock cod, the latter a valuable, beautiful and slow-reproducing fish that used to exist around the British dominion and old whaling base of South Georgia. By the early 1980s, the entire stock had collapsed.

Until the British declared a 320-km (200-mile) limit around South Georgia and the South Sandwich Islands in 1994 and began to enforce it with a new armed fisheries protection vessel, the *Dorada*, there was a toothfish free-for-all around South Georgia. After the British got a grip of that, using satellite surveillance and overflights from the Falklands to lead the *Dorada* to the poachers, the problem went elsewhere – to the poorly enforced 320-km (200-mile) zones around the sub-Antarctic and Antarctic islands under the jurisdiction of South Africa, France and Australia.

Even so, the first time many people heard of the hunt for toothfish poachers was when they heard the story, in August 2003, of the marathon 6,400-km (4,000-mile) pursuit of the Uruguayan-flagged toothfish vessel *Viarsa* by vessels of three nationalities. The *Viarsa* was spotted fishing near Heard Island and took flight. Believed to be carrying an illegal haul of toothfish worth £2.1 million, the vessel ignored repeated radio orders to stop and pressed on west, through mountainous seas, towards its home port of Montevideo, pursued by a lone, unarmed Australian customs ship, the *Southern Supporter*. For 14 days the customs ship ploughed on in pursuit of the toothfish pirates, through a blizzard and packs of icebergs, often losing visual bearings on the faster fishing vessel in huge seas and foul weather. It was later joined by a helicopter-equipped South African icebreaker, *Agulhas*, and by the South African salvage tug *John Ross*, manned by armed fisheries protection officers and capable of 30 kph (20 knots). It was not, however, until the ship was intercepted by the Falklands-based British vessel *Dorada*, equipped with a machine-gun and two fast inflatables, that the fleeing vessel was boarded and its 40 crew arrested.

What the pursuit of the *Viarsa* by an unarmed vessel showed is that Australian enforcement of its Antarctic fisheries was, until then, 'almost laughable', says David Carter of Austral Fisheries. The same, he believes is currently true of the French waters around Kerguelen Island and South African waters around Prince Edward Island, though there has recently been a spate of public criticism about sea bird deaths in French waters. Until recently, says Mr Carter, there has been 'simply a lack of political will, denying us adequate resources to defend our sovereign rights'. Mr Carter's company owns the right to catch 71 per cent of the annual legal quota of toothfish in Australian waters in perpetuity. He believes this arrangement has been instrumental in his company's vigorous defence of its rights. He and his fellow legal fishermen set up the Coalition of Legal Toothfish Operators (COLTO). In 18 months they have set up a website and published a rogues' gallery of the illegal vessels their boats see and photograph on the fishing grounds. He believes this would not have happened if they did not have property rights to defend. 'As a result of that interest, we have an enlightened self-interest in making sure it is protected. We have spent a lot of time, energy and passion in making this debate what it's been.'

COLTO's members say they have been distinctly unimpressed by the ability of the Hobart-based Commission for the Conservation of Antarctic Marine Living Resources (CCAMLR, pronounced 'kammelar') to take effective action. CCAMLR, set up by Antarctic Treaty nations in 1982 largely as a scientific management body, has many virtues on paper and in its treaty, which is one of the most enlightened where fisheries are concerned. (It is one of the few, for example, that explicitly recognises the need for 'ecosystem management', acknowledging that if Antarctic krill, which now have few ready markets, become overfished, the whole terrestrial ecosystem of seals and penguins could be faced with collapse.) For all CCAMLR's virtues, one of them is not enforcement. Being a consensus organisation on the UN model, like the equally pious and ineffective FAO, it declined COLTO's exhortation to blacklist persistent offenders. So COLTO took matters into its own hands and compiled the blacklist itself.

The rogues' gallery identifies at least two groups that have taken illegal fishing beyond the realm of shameless opportunism. *Viarsa* and its sister ship *Arvisa* were two of 26 vessels owned and operated by what COLTO calls the 'Galician Syndicate' based in Northern Spain. Many of the Galician Syndicate's fishing vessels were registered in Uruguay, others in Ghana, Argentina, Belize and Panama. Until the arrest of the *Viarsa*, there seems to have been great tolerance shown by Uruguayan officials to known poachers remaining on their registers, even generating valid CCAMLR catch documentation for them and accepting bogus satellite positioning data. Since the arrest, the *Viarsa* has been deregistered in Uruguay.

Flags of convenience are just part of the problem in trying to regulate the toothfish trade. Ports of convenience are a problem, too. Ports where toothfish vessels take on supplies include Port Louis in Mauritius, Durban in South Africa, Walvis Bay in Namibia and, of course, Montevideo in Uruguay. Since Mauritius, Mozambique and Namibia have all taken steps to tighten up controls on illegal landings, the poachers have taken to trans-shipping at sea. Once the toothfish has been transferred to a mother ship, as part of a huge cargo on its way to multiple ports in Asia, port state control becomes ineffective. This has been a major loophole in the Catch Documentation Scheme introduced by CCAMLR, which became binding in 2000, as has the dilatory behaviour of countries such as Canada and the EU in putting into force at home what they agreed to overseas.

The result of the *Viarsa* arrest, however, has been a political breakthrough for COLTO in Australia. Illegal fishing now has what David Carter calls 'taxi-driver recognition' as a major problem. Because its customs vessel was unarmed and could not make the vessel stop, the arrest of the *Viarsa* cost the Australian government AU$5 million (£2.8 million). Similar arrests are now seen as a false economy. Australia has, however, been getting serious about enforcement and copying the British example in South Georgia. The government has promised that AU$100 million (£40 million) will be spent over two years in patrolling the area with a new vessel armed with a 50-calibre machine-gun and capable of mounting armed boarding parties. France has also been getting serious about policing around the Kerguelen and Crozet Islands, where toothfish stocks are estimated to be only a quarter of what they once were. It has signed a treaty of cooperation with Australia, though this does not yet include mutual powers of arrest.

Meanwhile, however, the problem has just got bigger. Tagging studies have shown that 11 tagged toothfish did not stay conveniently within Australian jurisdiction, but swam some 1,600 km (1,000 miles) to Crozet Island, in French territory, across seas that are open for anyone to fish. This means the population outside the 320-km (200-mile) limits over which Australia and France have jurisdiction is likely to be significant and is in need of management. This in turn has highlighted the inadequacies of CCAMLR, which was set up as a scientific body, not a fisheries surveillance and conservation enforcement organisation. CCAMLR is all that stands in the way of the ongoing destruction of a significant toothfish fishery in the Ross Sea, off Antarctica. It is not enough. David Carter says: 'That's the powerlessness of CCAMLR. Ultimately, somebody's got to own those fish, otherwise it's the failure of the commons all over again.'

You need an ice-class vessel to fish in the Ross Sea, which means those Uruguayan and Korean vessels now fishing there in the Antarctic summer do so at some risk. How long will it be before someone constructs an ice-breaking fishing vessel, perhaps somewhere like Norway, where the government still dishes out shipbuilding subsidies, to exploit this beckoning opportunity?

CHAPTER 10
THE SLIME TRAIL

MERCAMADRID, 6.00AM. Madrid's modern fish market is the opposite of Tokyo's historic Tsukiji and, as a result, the envy of many visiting Japanese. It occupies 33 hectares (82 acres) of purpose-built covered halls and car parking off Madrid's N40 orbital motorway, and is serviced by roads, not by the sea. The first impression as you arrive is of a vast swarm of white vans belonging to the restaurants and shopkeepers of the city. Mercamadrid handles only 200,000 tons of fish a year, compared with Tsukiji's 600,000 tons, but that still makes it the largest fish market in Europe, the second largest in the world.

Although the Spanish capital is hundreds of miles from the coast, it is a fish magnet, and has been for centuries because of the Catholic monarchy. When the king was in the city, he and his court would observe the Church's formal requirement to eat fish at least two days a week (curiously, just what health experts now recommend). Madrid has long had an umbilical connection to the Galician ports. In the last century, when roads were less good, lorries racing to bring fish overnight to Madrid were often involved in spectacular crashes.

Mercamadrid is not particularly interested in introducing new tastes; it is in the business of satisfying the capital's traditional demands from whichever ocean has the right fish. You very quickly realise that there is an awful lot of hake on the market – surprising, given that scientists tell us that hake is on the verge of collapse in European waters. Perhaps it's less surprising when you realise that hake is a mainstay of traditional Spanish cooking. So the market is full of hake – great, gunmetal-grey fish up to 90 cm (3 ft) long, and one or two boxes of suspiciously small ones, at or below the minimum landing size, which are a delicacy. When you ask where all the hake is from, the answer is the same: 'The north of Spain.' This is the answer you get even if the hake is labelled as from Namibia, Argentina or West Africa and not actually from Europe at all. This answer does not betray ignorance but caution. One detects in the stallholders a reticence to look beyond the port where the fish is landed for fear of discovering a secret that might be better unknown.

Occasionally a case comes to court that brings to light the startling extent of illegality in the hake fishery, and the tolerance of rule-breaking in Spanish fisheries generally, evidence of regional cultures with a rapacious attitude to the sea and a striking disrespect for the conservation of stocks. There is northern hake, which is caught off Ireland and Scotland, and southern hake, which is caught in the Bay of Biscay. Southern hake is, if anything, in worse shape. But there's not much to choose between the two. There is very little European hake left. You do not need to look far to work out why: catches that are well over quota, and completely inadequate enforcement of those quotas in Spanish ports. In 2003, in a case that confirmed many of the suspicions British fishermen entertain about Spanish fishermen, an Anglo-Spanish fishing company was fined a record £1.1 million for a fraud in which its skippers caught more than 25 times the amount of hake they declared over two years, making £1 million in extra profits. The *Whitesands* was one of a fleet of Spanish-owned, British-registered vessels, based in Milford Haven, fishing off Ireland and to the west of Scotland and landing into northern Spanish ports, such as Coruña and Vigo. Swansea Crown Court heard that successive skippers of the *Whitesands* fiddled logbooks with 'breathtaking arrogance', declaring only 4 per cent of their true catches. Alterations downplayed the amount of hake the vessel had caught over 46 fishing trips by approximately 508 tons.

The Spanish company even carried on after the fiddle was uncovered and legal proceedings had begun. It then went into voluntary liquidation in a deliberate attempt to escape being prosecuted. Plymouth Shipping, which operated the trawler, and Santa Fe Shipping, which owned it, were convicted of 27 counts of fisheries fraud, including the alteration of logbooks, failure to record species under quota, and submission of incorrect landing declarations. To date no one has paid the fines. That same week a fine of £500,000 was imposed on the Spanish-owned, Milford Haven-based *Grampian Avenger* and *Grampian Avenger 2* for under-declaring 165 tons of hake.

The way the EU works is that the European Commission strives, not always that hard, to tighten enforcement against the sheet anchor of its member states, notably in the south. In a spirit of even-handedness, the commission recently issued proceedings against Spain and the United Kingdom for failing to enforce rules on fisheries. Commission inspectors found that Spain did not spend enough on its fisheries inspectorate to ensure the adequate inspection of fishing activities at sea or onshore. Lack of staff and lack of equipment were compounded by a confusion of roles between local and national authorities, incompatible electronic equipment and delays.

The commission noted failures to follow up observed cases of false declarations. It found that in the Canary Islands, where the vessels fishing in African waters land their catches, inspections were directed primarily at foreign vessels, while Spanish vessels were exempted. Shortcomings were observed in the inspection of vessels fishing under the agreement between the EU and Mauritania. On the mainland inspectors witnessed landings, sales and the transport of fish without any control by the local authorities. When they checked data recorded by local inspectors over that period, they found that only 15 per cent of the 200 tons of hake put on the market had been declared.

At La Toja, a classic Galician restaurant off the Plaza Mayor in Madrid, a waiter was laying out hake brought directly from Vigo in the refrigerated window. On the lunch menu were also cod, grouper, lobster, red bream, salmon, sea bass, sole, squid and turbot. House specialities included the classic *merluza alla gallega*, hake boiled with potatoes, garlic and peppers. They also serve hake neck – the head of a large hake cut in half and grilled – which non-Spaniards may not find to their taste.

Galician recipes are simple, not sophisticated, explains Manolo Sertage, the head waiter. They depend on the quality of their ingredients, so Madrid's top restaurants will pay that little bit more for the best. One has a sense that the clientele of La Toja will go on paying what it takes to eat what they have always eaten. At the heart of Spain, at the top of the food chain, everything rather spendidly remains the same. But a slime trail leads all the way from the sea to the people feasting in the city, something the feasters prefer not to think about.

'Black fish' – the name given to illegal catches – is Europe's guilty secret, not just Spain's, but Spain's part in it is the largest of all. I asked Harry Koster, head of the European Commission's fisheries inspectorate in Brussels, how bad the illegality was. He said that 60 per cent of hake, landed mostly in Spain, was unrecorded – in other words, illegal. How much cod in Britain was illegal? Fifty per cent. All official figures from the International Council for the Exploration of the Sea (ICES). All pretty disgraceful. Just as many of Europe's fish stocks are in danger of irreversible decline: every other cod, every other hake that hits your plate is stolen – from the general public and their grandchildren, the rightful owners of the sea. All because no one has the political courage to enforce the rules now that fish stocks are low.

One somehow does not expect the degree of criminality by fishermen in Europe's waters to be the same as that perpetrated by pirates in the seas around Antarctica, but it is, and that is still not readily understood when fishermen come on television with their tales of woe every time scientists propose a lower quota. The only difference between distant and inshore waters is that toothfish poachers make more money and do not have to go to the trouble of buying quota, licences, legitimate gear and observing rules on days at sea, which fishermen do in Europe before they can get down to fiddling. Toothfish pirates just happen to be cutting pieces off a, for now, larger cake.

The reason Spain is implicated so often in breaches of the rules is, one suspects, because of political pressures that have arisen because of its oversized fleet. This fleet fishes in more waters than any other European country. Unfortunately, what this means is that very often its boats break the rules, where they exist, on a global scale. Spain is not alone in breaking the rules outside EU waters. The Portuguese fishing fleet, though not as large, gives Spain a run for its money when it comes to illegality, particularly in the North Atlantic. The most spectacular example of what these two nations get up to is their flagrant illegal fishing on the edge of the Grand Banks, just outside the Canadian 320-km (200-mile) limit, some 12 years after fishing for cod and several other species was banned. A decade ago, Canada caused a major international incident when the fisheries minister, Brian Tobin, authorised the arrest of a vessel, the Spanish-flagged *Estai*, that was fishing illegally for Greenland halibut just outside Canadian waters. The EU accused him of acting illegally, since which Canada has tried to avoid restarting the dispute. It still has a controversial law on its statute book that bans boats flying flags of convenience from operating on the banks outside its 320-km (200-mile) limit, which is why vessels that fish outside that limit do so under EU flags, when previously they used a variety of foreign flags.

Ironically, one of the legacies of the *Estai* incident is that the Northwest Atlantic Fisheries Organisation (NAFO) now has some of the toughest rules of any regional fisheries organisation, intended to prevent the Canadians taking unilateral action again. These include the requirement that 100 per cent of vessels fishing in the area must carry independent observers at all times. The reports of observers on some of these vessels make chilling reading. They show that on 72 days of 2002, EU vessels deliberately targeted several species for which fishing was banned, primarily American plaice and cod. For some reason the detail of the observer reports is never published or made use of by the European Commission, except to check up on inspections by flag states. Only a bowdlerised version, missing many of the most damaging details so as not to cause international offence, is published annually in Canada. The observer reports simply seem to gather dust in a drawer in Brussels. So, using Europe's freedom of information legislation, I decided to ask the European Commission for reports on certain vessels. To my surprise they were duly produced.

It is vital to remember that we are not talking about an alleged crime of routine cheating or poor book-keeping, but a crime of deliberately steaming across an entire ocean to fish for stocks that are internationally recognised as having collapsed and on which there is a moratorium in the hope faint hope that they might one day recover. In economic and ecological terms, this is about as serious a crime as you get. It cheats not just present generations, but future ones, who might have hoped to benefit from the fish. The observer reports I obtained from the commission revealed that a territorial war of a particularly cynical nature is going on between fishermen and inspection vessels in the waters administered by the Northwest Atlantic Fisheries Organisation. Perhaps the most thought-provoking thing the reports show are the failures of the authorities in Portugal and Spain, and the apparent official tolerance of illegal fishing.

I read, for example, that the Portuguese-registered stern-trawler *Solsticio* entered NAFO waters on 19 March 2003. A British observer from the firm MacAlister Elliot & Partners was aboard. The vessel was watched carefully on its month-long trip by Canadian inspectors using the vessel's positioning data. The inspectors then boarded it on 5 May. They reported finding in the tunnel freezer, concealed behind empty racks of fish trays, evidence of previous tows, mostly consisting of the moratorium species cod and plaice. While the inspectors were on board, the vessel moved into deep water but continued to trawl, so when the net was retrieved, it contained 1.2 tons of fish, 60 per cent of which was Greenland halibut. This reduced the amount of cod and plaice to 40 per cent, just legal as a by-catch. The observer's figures showed that out a total of 284 tons of fish for the trip, 65 per cent was moratorium species: cod (83 tons), plaice (88 tons) and witch (deep-water sole, 14 tons). The vessel was reported to NAFO for 'directing' on moratorium species, i.e. fishing for them deliberately. From its satellite transponder records, it was identified as fishing for two weeks in shallow water where cod and plaice were found. It was reported catching moratorium species by an observer and found to have illegal species on board during an inspection (three sources, independent of each other). Yet when the vessel was inspected in port back in Portugal, the inspectors (one source) found there were 'no infringements' and said they found less than 10 per cent illegal species aboard. How much faith can one have in a system that allows such contradictory findings?

Two other Portuguese vessels, the *Calvão* and the *Lutador*, were reported for misreporting their catches and undertaking directed fishing of banned species in December 2002. Yet in both cases the Portuguese authorities chose not to recall the vessels to port for inspection. This lack of follow-up, say the Canadians, allowed both vessels to fish into a new quota year, thereby preventing identification of landed catch relevant to the 2002 inspection. The observer aboard the *Lutador* found that the vessel was catching and discarding small plaice while it was fishing for yellowtail flounder. The rules say that if a vessel catches more than 5 per cent of its haul as a banned species, it has to move 9 km (5 nautical miles). The observer's statement indicates that after 5 per cent of plaice were caught, the rest were simply discarded. An observer aboard the *Pescaberbes Dos*, a Spanish vessel, found quantities of undersized Greenland halibut caught on many occasions. 'Most of these fish were retained, and on a few occasions the skipper did not move the regulatory 5 nautical miles.' The observer, who is meant to weigh the catches himself, reported that he was asked to fill in his weekly reports of catches according to the figures he was given. Four other Portuguese vessels were accused by Canadian inspectors of fishing for moratorium species. Some were even photographed using tarpaulins to conceal illegal catches from air surveillance patrols.

EU vessels, of course, are not alone in misbehaving in NAFO waters. Canadian air surveillance has observed Russian vessels on numerous occasions with liners in the trawl, i.e. one net inside another, which gives a much smaller mesh size and catches greater numbers of juvenile fish. On 14 August 2003 Canadian inspectors approached the *Andrey Paskov*. As they approached, the vessel re-lowered the trawl under water. However, the cod end of the net floated to the surface and inspectors were able to see at close range a liner with an illegal mesh and a catch of 5 tons of redfish. The Russian vessel then released the entire trawl on to the ocean floor. The master of the *Andrey Paskov* was reported to NAFO for obstructing the mesh of his trawl.

It is difficult not to suspect someone in Brussels – but not those who made the information available to me – of conspiring to keep the flagrant and disgraceful behaviour of its vessels secret. Canadian officials find it frustrating to carry out inspections that are not acted upon by the flag state and to see observer reports that are simply filed by the EU. 'When you return to port in Portugal and Spain with large quantities of illegal fish, nothing happens to you,' said one Canadian official. 'There are 20 nations fishing in these waters and only two do not land in Canada. Which do you think those are?'

I put it to a member of the European Commission that there was plenty of evidence from observers to prosecute the masters of several EU vessels, even where the port inspector seems to have missed the evidence, so why did this not happen? The official said the reason was that observers' evidence could not be used on its own in court in Europe because it did not have any greater weight than that of the master of the vessel. It does in Canada, where the observer is seen as a dispassionate figure, whose evidence should, on balance, be believed. With European vessels, therefore, the burden of finding evidence rests with the port inspectors, who appear to be blind. Whether this anarchy will be improved by basing the new EU fisheries inspectorate in the Spanish port of Vigo remains to be seen.

Why is Spain so obviously the most ruthless fishing nation in Europe and possibly the world? Why is it also the most influential fishing nation in the EU? Partly because it has an oversized fishing fleet, as well as a distant-water fleet, when other countries, such as Britain, have given theirs up. Why does it have so many vessels? There lies a dark secret that more Europeans should be aware of the next time they tuck into a seafood paella on the quayside at Puerto Banus – or get pushed around by Spain in the Fisheries Council.

The expansion of the Spanish fleet coincides exactly with the time known euphemistically to modern Spaniards as the 'years of isolation', and to the rest of us as the era of General Franco. I am indebted to a paper called 'Spain and the Sea' by two academics from Seville University for explaining the connection. It points out that Spain had nourished grandiose overseas ambitions and stimulated shipbuilding in the early 20th century, but it was under General Franco's fascist government, which looked back approvingly to that time, that the fishing fleet underwent enormous growth. Fishing was the beneficiary of Franco's policy of protection and intervention, in which he favoured the navy, the merchant navy and shipbuilding. Spain's fishing fleet, in poor shape at the end of the Spanish Civil War, began to grow, and so did its catches – from 400,000 tons in 1940, the year after the Spanish Civil War, to 1,498,049 tons (a record never since surpassed) in 1974, the year before Franco's death. As a Spanish member of Greenpeace was the first to point out to me, joining the EU meant that the size of Spain's fishing fleet was officially supposed to be capped – one of several rules Spain's fishermen have been struggling against ever since. So it can be said that General Franco still has an outsized influence on Europe's fishing politics. No doubt he would approve, but do we?

Rules should ensure that European seas are managed commons. In fact, lack of enforcement and too many fishermen – Garrett Hardin's two definitions of management failure (see Chapter 9) – mean that Europe's seas are unmanaged commons in which the commoners and their fishy livestock face inevitable ruin. The growing contrast, I am beginning to think, is between those areas of the world where fisheries are commons, and those parts of the world where they are not.

* * *

Peterhead fish market, 7.00am, January 1997. The year Tony Blair became prime minister. Six thousand boxes of cod, haddock, monkfish, turbot, halibut, ling and saithe were being sold. I was there to investigate the extent of black fish landings at a time when nature had just delivered one last, miraculously large year-class of cod in the North Sea.

The port was promoting a new policy of high prices for fewer, high-quality fish, but this wasn't working. The harbour-master had asked the buyers and auctioneers, yet again, to wear white coats and not to walk on the boxes of fish. The buyers clustered around the auctioneer, and when they could not catch his eye, they clambered on the fish.

Slowly it became clear what we were seeing. The black fish was right before our eyes. Landings are measured in boxes, which at Peterhead should contain 57 kg (9 stone) of fish, but all of them were overfilled. They won't stack readily without crushing the fish, so must be reloaded by the buyer. An overfilled box, rather like a baker's dozen, theoretically meant a better price, and the extra volume went unrecorded because the boxes were not weighed.

Low prices were another indication of something afoot. It was the end of the month, quotas should have run out and fish supplies should have begun to dry up, but prices were flat. People admitted, quite openly, that lorry-loads of fish were landed in small ports around Scotland and driven to processing factories or wholesale markets in Hull or Grimsby. A 'private' sale, as it is euphemistically known, was agreed while the boat was still at sea. Fishermen would tell their friends where the inspectors were. Landing a few hundred boxes takes only a few minutes with a fork-lift.

'If the inspector should show up, the fishermen will simply declare the boxes he is landing,' one person in the industry told me. 'If the inspector does not appear, the skipper will unload his over-quota fish, then take the rest to one of the larger ports, such as Peterhead, and offload his legal catch.' Over time the skipper will keep spare quota, thanks to under-reporting, so that he can go on fishing when, in fact, his vessel should be back in port. That week I was in Scotland there were four landings at Montrose – 'that well-known fishing port', as one harbour-master observed dryly.

The fisheries inspectorate was poorly paid, subject to intimid-ation and had very little political support. When I was at Peterhead five years earlier, people used to deny there were black fish landings. But not now. 'There are very few people clean on the catching side in the industry now. Illegality has been institutionalised,' a board member of the government's Sea Fish Industry Authority told me.

The Scottish Fishermen's Federation admitted that their members were landing black fish. Bob Allen, chief executive at the time, explained that this was just the 'real world' his members lived in. One Scottish fisherman explained it simply: the cut in quotas that EU ministers had agreed made it necessary to break the law to stay in business. 'The way we have been regulated means that everything we were doing is now illegal. The ministry knew this would happen. My boat was built with an EU subsidy and was refitted with a ministry grant. We've got mortgages to pay. It's like telling British Airways that they can only fly to Paris once a day when they have been used to flying six times.'

A senior inspector then told me that 50 per cent of the cod and saithe landed was illegal. Lord Selborne, chairman of a House of Lords select committee, said it was 40 per cent. Whatever the figure, the scale of the landings overwhelmed the 12 per cent cuts in fishing quotas for cod agreed that year. Over the next couple of years the huge year-class of juvenile cod was squandered, mostly as by-catch.

Meanwhile, the black fish was on the lorry and heading south to where the processors are, in Hull or Grimsby, ports for the now almost defunct long-distance fleet, which has now virtually given up catching fish. Documentation was, and still is, easy to falsify. If a lorry is stopped, there is no way of telling where the fish are from, and the inspectorate does not consider it cost-effective to monitor the fish after the first sale.

Large companies involved in fish processing sell to the super-markets, so the fish one eats, if it comes from the North Sea, stands a 50–50 chance of being 'black'. These big companies have shareholders who are not used to dealing in illegally obtained goods, so questions have been asked. The directors of Booker Fish, the largest fish processor in Britain, sought legal advice on whether it was an offence, knowingly or otherwise, to buy illegally caught fish. Their lawyers said that the offence was catching it, not buying it, unless collusion could be proved.

Another processor said: 'You get a call saying, "Do you want 200 boxes of haddock?" It arrives by lorry. It goes into the major processors. They won't know that it's illegally caught. They won't ask where it is from. They'll ask what the quality is and when it's going to arrive.'

Some of the more correct firms could not cope with the fiddling that came over the industry when the quotas, aimed at saving fish for the future, began to bite. Irvines of Aberdeen told *Fishing News* that black fish was one reason for their decision to quit the business. Some large firms had been told by their accountants that they were not viable without black landings. Half the fishermen in Peterhead are teetotal Presbyterians; the saying goes that the other half drink twice as much to make up for them. As one Grimsby merchant put it: 'I've got friends in Peterhead who wouldn't buy fish caught on a Sunday. But they have a responsibility to their staff and they have had to become complicit with black fish.' Very few would have nothing to do with it. Only one processor on the docks at Peterhead, a member of the Free Kirk of Scotland, was said to refuse to handle black fish. The understanding at that time was that nearly everyone else was willing to.

Although fish quotas were being cut, boat-building on the east of Scotland was booming. Boats theoretically designed to catch deep-water species to the west of Scotland were being turned out of Scottish yards. What they were actually going to catch was black fish. Other boats were being built with little quota to fish on. Bank managers were aware of this quandary. One asked a skipper just how he intended to finance his boat from his modest quota and was told it would be from black fish.

David John Forman was chief executive of the largest fish

processing company in Peterhead. Outside his office window an army of women in sterilised overalls and hats were filleting fish. A cutting on his wall recalled how he pulled the crew of a Danish trawler from their sinking vessel in a force ten gale. He recalled the moment vividly and described how, as he arrived at the wreck, the lights of the trawler were already shining under the sea.

I asked him about black fish. He replied: 'The quotas are uneconomic.' His meaning was plain: skippers would go bust if they stuck to their quotas – and nobody was going bust. In his view: 'The government has been quite tolerant.'

It was, after all, an election year.

That was seven years ago. When I published most of the above in *The Daily Telegraph*, there was an outcry. Many British people had been brought up on the great myth of the commons – that only foreigners cheat. The news hit their ideas hard – for a day or two – until they got used to it. There was a crackdown of sorts by the newly installed Labour government, which insisted that fishermen land at designated ports, as they did in Holland. My contacts told me not to come to Peterhead or I would be beaten up.

Tony Blair is still in power as this book goes to press. There are fewer cod in the North Sea, and every other cod landed is illegal once again. About a thousand fishermen left the industry last year. The trail of slime from black fish continues to touch all levels in the fishing industry, together with the banks who lend to fishermen, the food processors, the supermarkets and even the consumer, who benefits from continually flat prices.

A significant change that *has* come about is Scottish devolution, which has made Scottish politics more like those in Spain. Now fisheries are run from Edinburgh as well as London. The Scottish Parliament is even more responsive to the plight of the fishermen, rather than the fish, while the London-based 'English' media, which is often resented in Scotland, sees things the other way round.

There seems to have been precious little political courage shown by the politicians of devolved Scotland, very little attempt to stop the further squandering of a common resource in the face of determined campaigning from fishermen and by the Cod Crusaders, two fishermen's wives from Fraserburgh, Carol MacDonald and Morag Ritchie, with their slogan 'Save our cod'.

In a splendid exchange before a House of Lords select committee, Lord Selborne asked Elliot Morley, the London-based fisheries minister: 'So save our cod means kill our cod?'

'Yes,' replied Morley.

In its legal letter of complaint to the UK, once a more law-abiding nation, the European Commission noted a failure by the authorities to cross-check data and a failure to take action against fishermen who broke the rules on a number of occasions. The commission noted cases of misreporting, where vessels recorded catching fish in one area while their 'spy in the sky' satellite records showed that they were in another. To avoid detection, vessels often turned off their blue boxes altogether. This did not seem to be reported or penalised by the authorities, who, despite much fanfare in the 1990s when they were introduced, did not appear to use the blue box as an enforcement tool.

Fishermen got up to tricks once thought of as Spanish: concealing cod in large containers labelled ling, and misreporting saithe, cod, hake, megrim and monkfish as ling, greater forkbeard, tusk and dogfish.

Mr Blair, who had done almost nothing in the previous seven years for the fish stocks around Britain's shores, except eating fish in preference to beef at official functions, set up an inquiry in 2003 into the fishing industry. Written by the Prime Minister's Strategy Unit, a think-tank run by some of the brightest civil servants, the inquiry's report turned out to be a thoughtful and comprehensive piece of work, which told the industry to reform or die, but in relentlessly upbeat, positive terms amusingly reminiscent of Mr Blair's comic *alter ego,* the Vicar of St Albion in *Private Eye.* It identified lack of compliance and the excessive size of the whitefish (cod, haddock and plaice) fleet as the 'two major challenges to sustainability and profitability'. The changes it recommended to stop black fish landings – administrative penalties, such as the loss of a licence for infringements, electronic logbooks linked to markets, on-board observers and improvements to the on-shore paper trail – were sensible enough. These have yet to be accepted or adopted at the time of writing.

The whole food industry – the sellers, the buyers in supermarkets and processing firms as well as the fishermen – bears responsibility for what has been going on in the sea whether or not they are aware of it. Given the scale of illegal catches, it is virtually impossible for anyone to put their hand on their heart and say they have not, inadvertently, bought, sold or eaten, illegally caught fish. No one can back away from that, not the humblest fish and chip shop nor the most famous restaurant in the land.

CHAPTER 11
DINING WITH NOBU, MARTHA, GORDON AND NIGELLA

Humans are a virus.

Agent Smith, *The Matrix*

THE CULINARY establishment says that we should eat more fish. Delia Smith, Britain's best-known domestic diva, and her US counterpart, Martha Stewart, exhort their television audiences to do so – and from a health point of view they are absolutely right. It would do many of us good to eat more fat-free protein and more oily fish rich in omega-3 essential fatty acids that improve heart and brain function and even, some believe, reduce hyperactivity in children. Delia, whose dependable dishes have been the saving of many dinner parties and more than one marriage, is merely repeating the prevailing view, backed by the British Heart Foundation and similar bodies throughout the world, that we should all try to eat at least two portions of fish a week, at least one of portion of which should be oily fish. That view, and the growing scarcity of some fish, has boosted the fish price in Europe, the USA and Australasia to the point where fish is no longer the food of the poor – it is the food of the health-conscious chattering classes. However, if we want to be informed and thoughtful human beings ruled neither by instinct nor fashion, we must look at all the facts and conclude that the unqualified advice to eat more fish is already past its sell-by date.

Any civilised society concerned for future generations must pay attention to sustainability and attempt to favour the 'right' fish, namely, those caught without causing them, or any other species, irrevocable harm. If that attempt is not made, we are going to run out of many species of wild fish and be left with only farmed ones. Unfortunately, farmed fish have all the problems of drug traces, pesticide use and feed contaminants – which have caused a stampede away from intensively farmed meat and towards locally grown, extensively reared or organically produced livestock on land. That is why wild fish conservation is a human health issue.

For the sake not only of the world's wild fish, but of future generations who might like to eat them, one might think there were now very good reasons to start eating *less* fish, at least of certain kinds. This is a drastic measure, but the trends tell us all too clearly that some of us are eating too much. A more preferable alternative to eating less fish – though this is seldom possible because of a lack of both information and choice – is to go on eating the same amount of fish less wastefully caught, i.e. caught leaving more small fish and other animals in the sea. If we do that, we might some day end up eating fewer but larger fish. But how do you start that process, other than by starting unjustified food scares? Well, one possibility is to suggest ways of eating just as healthily, while eating less fish. This could be achieved by eating fresh fish of an even higher quality with all their omega-3 fatty acids intact. These oils are found in fresh herring, mackerel and tuna, but not in canned tuna because the oils tend to be lost and replaced by soya, olive or sunflower oils in the canning process. The canning market is very selective about which chunks it uses, discarding all 'black meat'. This and other waste fish goes to feed cats and dogs. Is this wasteful, with the petfood industry a hidden factor behind our rising tuna consumption, or is it a sensible use of a waste product for which there is no other use? To choose fish caught less wastefully, the consumer is going to need much more information.

Eating less fish, or fish caught less wastefully, may be what we should be doing, but this message does not appear to have got through to some of the leading lights of the culinary world. In the world's top restaurants and hotel kitchens, chefs who know to flatter their customers with slimming, healthy and exclusive foods seem not to have heard that some fish may no longer be a renewable resource. At the aspirational end of cuisine, some of the world's leading opinion-formers appear not to ask themselves very often whether, if they go on serving the dishes they create and describe so lovingly in their recipe books, there will actually be anything to serve their customers in five, ten, 20 or 30 years' time. A lack of curiosity about how fish are caught and whether there are enough of them to justify eating them seems endemic among most top restaurateurs and food writers – with some honourable exceptions – as well as among less exalted retailers. At the very least, it must be said that some obvious opportunities to inform customers about the effect fisheries are having on the world and to shape the path of sustainable consumption are being missed.

Singling out any individual chef or writer may seem unfair when it is a convention across virtually the whole world in cookery books and on menus to mention very little about how fish are caught and whether one should in all conscience eat them. One has to ask, though, whether this convention is in itself sustainable. And, in looking at current practice, one has to start somewhere. The place I think we have to start is with the chefs who have taken the highest profile in forming new trends in the consumption of fish and in suggesting new ways of eating them to new audiences around the world.

Though I could have chosen dozens of other names in the culinary world, Nobuyuki Matsuhisa stands out as an example of a brilliantly inventive exponent of Japanese and 'fusion' cooking and an ambassador of Japanese culinary tastes and techniques to the world. His 13 Nobu restaurants are the haunts of film stars, supermodels and tennis stars on three continents. Nobu was trained in Japan and cooked in Peru, Argentina and Alaska before opening a restaurant in Beverly Hills into which Robert De Niro, actor and part-time restaurant entrepreneur, one day happened to walk. The friendship and business partnership they formed is one of the greatest success stories in the restaurant business. Nobu is known for creating imaginative, expensive and stunningly beautiful dishes that fuse the pure, often bland tastes of sushi and traditional Japanese cooking with the North and South American flavours of garlic, chilli and coriander. He is probably even better known for the clientele of his restaurants, where the reservation list can read like a night at the Oscars: Gwyneth Paltrow, Renee Zellwegger, Nicole Kidman, George Clooney . . . Only last year, in a survey, the Nobu restaurant in London, a favourite of the late Diana, Princess of Wales, knocked The Ivy off pole position as London's most fashionable restaurant.

Nobu has become a master of publicity, as well as of sushi and sashimi, and he has used this mastery to expand the market for his kind of Japanese food. He teamed up recently with Martha Stewart and Robert De Niro to write his first cookbook. (Martha Stewart, incidentally, made a webcast with Nobu in Tsukiji fish market in Tokyo, enthusing over all those beautiful tuna, before she mounted her own invasion of the Japanese market with her particular brand of cooking and decorating accessories.) Before you open the covers you cannot fail to notice that *Nobu: The Cookbook* has more celebrity endorsements than any book, let alone a cookbook, you have read. Words of praise from Madonna, Giorgio Armani, Bill Clinton, Andre Agassi, Robin Williams, Cindy Crawford, Leonardo di Caprio and others jostle for space on the back cover. Kate Winslet, for instance, says: 'The food at Nobu can be described in two ways – quite simply heaven on earth and SEX on a plate.' Stephen Spielberg says: 'Your food is the BEST. Just don't tell my mother.'

Such endorsements mean one can hardly wait to eat one of Nobu's specialities – for instance his 'New style sashimi', a di Caprio favourite, which looks particularly innovative *and* delicious, being partially cooked in drizzled hot olive oil. The list of fish Nobu chooses to cook may be sublime in terms of quality, exclusivity and taste, but if you know the current state of fish stocks, you have to say he is vulnerable to criticism. Check out the recipes in his cookbook against the prevailing ecological advice on what not to eat, set out, for example, in the Seafood Watch Program website run by the Monterey Bay Aquarium in California, the guide published by Blue Ocean Institute and the Marine Conservation

Society's *Good Fish Guide*, written by Bernadette Clarke and published in the UK. In Nobu's cookbook you will find many species that are in very poor shape, such as abalone, Caspian caviar, Chilean sea bass (known outside the USA as Patagonian toothfish), grouper (a coral reef fish), red snapper (sometimes fished with dynamite), sole and the finest sashimi-quality tuna steaks. In his book Nobu makes no reference to which tuna he uses, describing all tuna as simply *toro*, or tuna. He does, however, say he prides himself on the very finest quality ingredients so I wondered whether he was using bluefin or bigeye: the most sought-after tuna. I contacted his restaurant in London to ask what tuna he used and what was their sourcing policy. Sadly, they did not respond to requests for information on the subject. His cookbook also includes exotics such as flying fish roe (which does make you wonder what happens to the rest of the flying fish) and the notorious blowfish or fugu. All but the last two species mentioned so far have known problems of sustainability in one ocean or another – which means we may be killing it off – with the exception of the fugu, which can kill *you*. While lethalness to humans is something Nobu obligingly draws attention to, he is not one of those cookery writers who mentions anything about the lethalness to fish species of any fishing methods or management regime. He doesn't take the opportunity to help you to eat less fish or to eat the same fish less wastefully and that, I believe, is a shame. He does, for instance, say that some abalone – the subject of homicidal poaching by rival gangs in South Africa and of a total ban on harvesting in California – is now farmed, but not how you can tell one from the other.

On the West Coast of the United States, where the 'Take a pass on Chilean sea bass' campaign has altered the menus in many restaurants, publishing recipes for a fish as controversial as the Patagonian toothfish, as Nobu does, is bound to be remarked upon. There are, of course, some complex ethical issues about toothfish, which environmentalists, in their hurry to be pious, don't often tell you, and which Nobu might have taken the opportunity to mention in justification of his decision to serve toothfish. A simple-minded boycott, as advocated by US environmental groups, devalues not only the fish caught by trans-oceanic poaching syndicates but also the legally caught Australian toothfish and South Georgia toothfish, which is now certified as sustainably managed. But how do you know the fish on your plate is legal or sustainably caught? Nobu doesn't say where his toothfish comes from. When I called his people and asked about his fish purchasing policy, an email came, apologising for the delay, then no follow-up. So in the absence of a 'chain of custody' to identify legally caught or sustainably caught toothfish, I am left having to assume that Nobu doesn't know where his toothfish come from any more than the rest of us. His Chilean sea bass and truffles with yuzu soy butter sauce may, in other words, just be plain old poached toothfish.

*   *   *

Gordon Ramsay is Britain's top chef, his website tells. Whether or not you accept him at his own assessment, which is engagingly devoid of false modesty, you have to accept him at something like it, because Ramsay is very, very good indeed. And an area of territory he has marked out for himself is the cooking of fish, not least by posing naked on the cover of one of the Sunday magazines with nothing but a giant flatfish to preserve his modesty. Trained in France, Ramsay cooks with the simplest and best ingredients, creating dishes with a flair and attention to detail that make reviewers go weak at the knees. He has been rewarded with three Michelin stars at his Chelsea restaurant and his protégés cook at Claridges, the Savoy Grill, Petrus and the Connaught. Ramsay has also produced his own cookery book about fish, *Passion for Seafood,* co-written with Roz Denny. This mouth-watering read by a chef of formidable imagination and technical skill, however, includes some notable contradictions.

The passion is real, indeed infectious. Ramsay grew up in Glasgow and learned his love of the sea in the west of Scotland. He reminisces about eating winkles with a pin in his native Glasgow and fishing for salmon with a worm on a squeegee bottle float on Loch Lomond or with a fly on the Tay. Then he talks of expanding his knowledge of fish on a luxury yacht in the Caribbean, learning to dive and finding out about the habits of some of the beautiful creatures to be found beneath the waves. He writes: 'The manta rays were the most magnificent as they wafted past with their billowing fins.' Remember those manta rays with their billowing fins. Ramsay concludes his signed introduction with all the campaigning ardour of a conservationist: 'All these experiences have left me with a lasting impression of the need to protect the wealth of the sea, from the powerful North Atlantic with its dwindling stocks of cod and white fish to the fragile, dreamlike beauty of the coral reefs.'

Then the contradictions begin. For while Ramsay's cookbook does more to help the consumer make informed choices about fish that are sustainably caught or farmed than Nobu's does, it still recommends fish that the best ecological advice would be to avoid – because it is seriously overexploited. His surprising advice on tuna, for example, is this: 'I buy line-caught bluefin (the best) or sometimes yellowfin.' Now, no one would criticise him for recommending yellowfin, still one of the most plentiful tunas. Bluefin is a different matter. The *western* Atlantic spawning stock of the world's most valuable fish has been hammered to around a tenth of its level in 1970. No responsible fishery manager, in the view of most conservationists, ought to allow a stock to decline that far without imposing drastic curbs on fishing, or better still a total ban for several years to allow the population to recover. But ICCAT has stood by for 30 years while the decline has gone on without taking such action. The *eastern* Atlantic population of the bluefin tuna (and we now know the two intermingle more than was previously thought) is now a cause of concern to ICCAT scientists because of ludicrous quotas that the EU and other members of the Commission have awarded themselves to satisfy the demand for wild stock for tuna 'farms' in the Mediterranean. Even before this surge in fishing effort, the bluefin tuna was included in the International Union for the Conservation of Nature (IUCN) Red List of endangered creatures – along with the black rhino, the elephant and the great apes. Globally its stock is listed as 'data deficient' because little is known about stocks in the Pacific, but in the eastern Atlantic bluefin is officially known as 'endangered' and in the western Atlantic as 'critically endangered'. The seafood guides I have mentioned say it is one to avoid, and I for one would not disagree.

No one in a position of influence, in my opinion, ought to be buying, or endorsing the buying on a regular basis of bluefin tuna, let alone someone who tells us, as does Gordon Ramsay, that he is passionately convinced of 'the need to protect the wealth of the sea'. How it is caught makes no difference. I am not sure what Ramsay is trying to say by telling us he only buys 'line-caught' bluefin. He might mean that the fish is unmarked by nets and therefore of the highest quality. But what sort of line does he mean it was caught on? Does he mean caught on rod and line by big game fishermen off the coast of the United States? These sport fishermen, who sell their catch to Japan, managed to catch an impressive 2,070 tons of bluefin out of a total of 3,215 tons caught on rod and line in the Atlantic 2002. Or does he mean bluefin caught on the commercial long-lines set elsewhere in the Atlantic by Japanese and other fleets, with, no doubt, a considerable by-catch of other species? Long lines, which are often dozens of miles long, accounted for 4,920 tons of bluefin in the east Atlantic and 727 tons in the west in 2002. It makes not much difference to the bluefin's eventual survival whether it is overfished by either method. The fact is, it is overfished. The bluefin is, in every sense, at the end of the line.

There is an argument that some catches, properly regulated, are better than none for practical reasons. Legal fishermen, as salmon fishermen have found, help to police the sea and prevent poaching. Some legal big-game fishing might be sustainable if one could be sure that catches were low enough, but the fact is that most people in the conservation world have lost confidence in ICCAT's stewardship of the bluefin, to say the least. Some say it is a disgrace. So there is a compelling argument that bluefin ought by now to be protected by stringent quotas and rigorous controls on trade – as it would be under CITES. And it would be listed under CITES already, but for some disgraceful politicking by the United States and Japan in 1992, when Sweden proposed protecting it and was humiliatingly rebuffed. The result, more than a decade later, is that this amazing creature is fast joining the great whales in the ante-room to extinction. Somebody really should tell Gordon Ramsay this, for the bluefin badly needs a high-profile champion like him.

Perhaps, I thought, I should actually make that suggestion. I tried to contact Mr Ramsay three times on the telephone to ask him about his fish sourcing policy in general and his use of bluefin in particular – without success. The third time, someone from Gordon Ramsay Holdings left a message on my answering machine saying: 'On this occasion Gordon Ramsay will not be able to participate in your book. With every good wish and thank you for calling.'

This was a pity, as I would have liked to discuss with Gordon Ramsay the particular dilemma he shares with the *crème de la crème* of the world's restaurants of being expected to provide hard-to-get and expensive fish. This increasingly means the most endangered. His fish book has recipes for halibut (this is the fish he is holding over his private parts on the cover of the Style section of *The Sunday Times*). Halibut is officially listed by the IUCN Species Survival Commission as globally 'endangered'; that is, it stands a 'very high risk of extinction in the wild in the near future'. Ramsay also has recipes for turbot, cod (though he does warn that this is overfished), gilt-head bream (a fish prized by the Senegalese but increasingly rare in their waters), snapper, tiger prawns and sturgeon. He does point out that the gilt-head bream and the sturgeon are now often farmed, though one might add that this does cause other ecological problems. Ramsay also includes a large and knowledgeable section on caviar. 'It may be the ultimate luxury food, but caviar is being eaten by more and more people,' he writes without any apparent sense of irony.

Couldn't he try just a little bit harder than that? More and more people eating caviar is precisely what has put every species of sturgeon on the endangered list. Beluga, Ramsay's top recom-mendation, is the most threatened species of all. Things look about as bad as they can get for the sturgeon. Sturgeon take 17 years to reach sexual maturity and are killed in the process of extracting their eggs. The poaching of wild sturgeon in the countries around the Caspian Sea has been rampant since the fall of the Soviet Union and dominated by mafia gangs. There remains a huge sustainability problem with caviar. Traffic, the wildlife investigation arm of the WWF, recently discovered that there was a huge illegal Russian market in caviar, accounting for something like 80 per cent of domestic sales – which western consumers can do very little about. Some poachers have actually been trawling the Caspian for sturgeon. But – as Gordon Ramsay and other chefs might justifiably point out – boycotting it deprives it of value and is not necessarily the answer to ensuring the sturgeons' survival. From 1998, the trade in caviar came under regulation by CITES, which has brought some improvements, but the process of conserving the sturgeon is still hampered by some vastly over-optimistic stock assessments originating from, not surprisingly, Russia. Gordon Ramsay says he buys his caviar from Iran, where caviar is a government monopoly and under better control than in other Caspian-range states. This is good advice. The danger comes in buying caviar out of a suitcase, without the CITES-approved label the jar is now meant by law to carry.

Caviar is undoubtedly a difficult one. The trade which reaches the west is relatively well regulated but there are huge illegal pressures on the sturgeon. Given the scale of the problem, some have decided to give the sturgeon a break until the situation in the Caspian improves. The occupants of the Michelin guide are, to be fair, caught in a quandary, for the loucher kind of customer at the world's top restaurants has always wanted something no one else can have. To some, rarity is just another form of exclusivity. It is a dilemma that the rest of us will never have to face, not having to knock out 'Eggs of endangered fish with spaghetti and crème fraîche' at, say, 20 covers a sitting. But you do feel some could try a bit harder to make what is *plentiful* more desirable.

Whether it is the chefs or their customers who are to blame for creating these expectations of biological exclusivity, there are some chefs and cookery writers who feel uncomfortable about them.

Richard Whittington, author of many books, and for three years culinary consultant to Terence Conran's chain of restaurants, says that this is part of the appeal of many top restaurants. 'You pay a premium for the privilege of knowing that you are eating something that is about to go extinct,' he says. When you put it like that, it becomes difficult to say who is more to blame for pushing fish towards extinction, the chefs, or their customers.

Whittington decided he wanted to make a stand. He persuaded Conran to take cod off the menu in protest at the overfishing going on in the North Sea, until Icelanders persuaded him to put it back on. 'I wanted to draw attention,' he says, 'to the criminal fishing in the North Sea and make chefs look at alternatives and become aware of when it was more appropriate to eat certain fish because of spawning times.' Whittington is passionate about putting more information on the menu. It has already arrived with meat, he says. It is coming with fish. Whittington is planning a restaurant called Provenance, and says it will offer carefully sourced and labelled produce, both meat and fish.

*   *   *

While Cornwall-based restaurateur Rick Stein has no Michelin stars, he does have a strong, national following. To me, his first TV series about cooking with fish came embarrassingly close to saying 'Eat fish while stocks last'. But Stein then became more thoughtful about the ethical quandaries of which fish to eat and gave his time for nothing to charities seeking to promote the choice of sustainably caught fish. His later programmes have recommended little-cooked species to give the common commercial species, such as cod, haddock and plaice, a break. Stein's your man if you want to know how to cook a by-catch fish: he does excellent recipes for gurnard. When I spoke to him on the telephone, this was his view of his colleagues:

Most chefs have got very limited knowledge of where their supplies are coming from. Their job is locating stuff. A few of us think the source of supply is as important as the cooking. It may be a slightly self-serving point of view, but I've always said that unless we become more enthusiastic about fish and therefore more knowledgeable, nobody is going to conserve it.

Mmm, I may have said, there are those who would see that as a *bit* self-serving. Stein sounded hesitant. It is a quandary putting together a menu, he said, when people come in and say should cod really be on it? He felt he could defend cod because it was in a better state around the Cornish coast than in the North Sea or the west of Scotland, and most of his fish came from the more sustainable, inshore boats. He is a supporter of no-take zones, where no one can fish and stocks can recover. Stein admits he doesn't have all the answers, but at least, like Richard Whittington, he is a pundit who shows evidence of having thought about the options open to him. The chefs who pretend the problem doesn't exist are as disappointing as a bad meal in a pricey restaurant.

That brings me to the clientele of the world's pricey restaurants. I'm curious. Do the likes of Cindy Crawford, Bill Clinton and Andre Agassi not ask where their fish is from and how it is caught? Have they even thought of doing so, or do they believe that the international standing of the chef whose food they are eating means that it must be all right? Or does their own desire to eat healthily simply come before asking difficult questions? If so, they are not alone. If you type omega-3 fatty acids into Google, you get links to over three million documents. For some people, what goes into their bodies has become an overruling obsession and you can cruise the world's Omega-3 websites without encountering any reflections about where these prized fatty acids are coming from and at what social or environmental cost. Could we be identifying a successor to the Me Generation – namely, the Don't Care About the Rest of the World As Long As I Have a Spa and Some Omega-3 Fatty Acids Generation? Let's call it the Omega-3 Generation for short. Personally, I would rather believe that celebs just don't know much about fish. At least not yet. Having found out a little more, these are the sort of people who will probably ask whether there's any eco-friendly stuff on the menu, and if not, they'll be out of the latest trendy restaurant faster than you can say 'politically correct'.

It's all very well to make fun of the celebs being ushered to their reserved seats in the smartest restaurants, but we ordinary mortals are implicated in this too. Those of us who cook out of fashionable books by the likes of Jamie Oliver and Nigella Lawson are in on the global trends behind the growth of Mediterranean and Japanese food – both of which cuisines have arguably enlarged their appeal for health reasons. The result of these trends is that overfishing is expanding into the high seas and the ocean depths, the last refuges of healthy populations of fish. Some of the wealthy seem to be celebrating Nero-style. Shame on them. But we, too, need to take another look at what's in the freezer, the refrigerator and the can on the shelf.

Tuna are one of the theme fish of Mediterranean cuisine, but that sea has itself long been empty of anything approaching a middle-aged fish. Tuna are the theme fish, too, of Japanese cuisine, which vies, neck and neck with the EU's much larger population, for the lead in emptying the oceans of the world. Fresh tuna is a new entrant in recipe books in northern Europe, as is swordfish. Nigella Lawson, a cook with many influences, has recipes for tuna carpaccio and swordfish steaks. A generation ago, she would only have been writing recipes for sole and lobster. Come to that, look at the recipes in women's magazines. Those who used to eat fish and chips are increasingly eating Mediterranean fare.

In Europe this has meant an upward surge in sales of tuna, to the point where it is now the biggest market in the world for sales of canned tuna, just ahead of the United States. It's a popular option as an everyday lunch, and few store cupboards in the developed world do not have a couple of cans of tuna in them, waiting to be used in a quick pasta bake or a salad Niçoise. But do any of us know how that tuna is caught? And what would we do if we did know?

As I began this book, I carried out a quick test on my own household by rummaging along the kitchen shelves. I found four cans of tuna steaks or chunks. Three were 'own brands' produced for the supermarkets Tesco, Sainsbury's and the Co-Op, and one was from John West, the leading Heinz brand. Only the last of these actually said which species of tuna it was – skipjack – on the front of the can. Sainsbury's gave the information on the contents list, but Tesco and the Co-Op didn't give it anywhere – even though they tell you how much energy, fat, protein, carbohydrate, salt and other additives the can contains. Not that they are required to. Regulations governing the labelling of *fresh* fish in the year 2000 do require the fish to be labelled with its commercial name, whether caught or farmed, and the area where it was caught, to be provided at retail sale. Sadly, canning regulations haven't caught up. EU regulations, dating back to 1992, just require the industry to specify 'tuna', not which species. If this tells you that tuna is less often seen as a living creature than as a global commodity, you would be right. Tuna migrate across whole oceans and are flown dead around the world, making the planet simultaneously smaller and more polluted. One of the ways of feeling less concerned about this and the fellow creatures you happen to be consuming is not giving them a name. The cans did tell you, as they are now required to do, the place of origin, which was truly global: Mauritius, Thailand, the Maldives and Ecuador. The Co-Op called back to say they now carry this information but I might have been looking at one of their old cans. Did I know that the shelf-life for tuna in oil was five years, and three years for tuna in brine?

What I found particularly surprising was that although half the tuna canners didn't seem bothered enough to tell you what kind of fish was in the tin, they all wanted to tell you that their tuna was 'dolphin-safe' or 'dolphin friendly'. They actually thought it was more important to tell you that than what species of fish you were eating. This seemed to me to be evidence of the same kind of myopic, mind-warping 'speciesism' that I have identified before in mammal-obsessed animal welfare groups. Look up animals in the dictionary – fish are included.

So what does this 'dolphin-friendly' piece of information mean and why should it be so important to fish canners that they want to tell us about it? I did dimly remember, as a layman, that dolphin by-catch was a problem for the catchers of yellowfin in the eastern

Pacific, but I had read that there were other by-catches, such as turtles and sea birds. Hadn't the British Antarctic Survey just said that the problem for albatrosses was no longer the toothfish boats but the Asian tuna long-line fleet in the Indian Ocean? So, I wondered, why didn't they tell us it was turtle-friendly tuna, or sea bird-friendly tuna, or shark-friendly tuna, or even tuna-friendly tuna (caught in a way that means there will always be more tuna)? The reason, I have now discovered, for not telling us that, is probably because the way they catch the tuna that goes into cans is not friendly to very much at all.

I phoned and asked the spokeswoman from Sainsbury's, a store that supports such green initiatives as the Marine Stewardship Council, whether there were any sustainability problems about the tuna in cans. 'None,' she said. This appeared on the face of it to be a bit too absolute to be believable, so off I went to find the answer for myself.

Tuna canned for Europe and the United States is caught by purse-seiners. The FAO estimates the size of this fleet at 570 vessels worldwide, ranging from 250 to 4,000 tons. Spain and France are big tuna-fishing countries, especially in the Indian Ocean, the Atlantic and the Pacific. The USA, Mexico and Ecuador are big operators in the eastern Pacific, and Australia is a big catcher in the Pacific and the Indian Ocean. Tuna eaten fresh as steaks, sushi or sashimi comes from the long-line fleets of countries such as Japan and Taiwan, whose baited hooks and lines up to 130 km (80 miles) long cause a significant by-catch problem of albatrosses, endangered turtles and sharks. Nobody knows how many long-liners there are, but it is a lot. More worryingly, nobody appears to be counting. The other significant way of catching tuna, particularly around the islands of the Indian Ocean, is by using pole and line from bait-boats. Bait-boats attempt to create a feeding frenzy by throwing chopped-up bait at swimming schools of small tuna, often using fire-hoses to whip up the surface water. Then individual tuna are hauled out by pole and line. The size of the tuna caught in this way tends to be small, but there is virtually no by-catch. The resulting tuna supplies the growing fresh tuna market in

Europe. This pole and line method, also used to catch albacore in the Atlantic, is the least-bad method of fishing, and worth remembering when you buy.

The targets of the world's purse-seine fleet are the tropical skipjack and yellowfin tunas. The 90-cm (3-ft) skipjack is incredibly fecund, with a short life and a high natural mortality rate. It was described to me by Dr Geoff Kirkwood of Imperial College, London, as the rat of the sea. I've also heard it described as the cockroach. Like mackerel, to which it is related, the stripy skipjack constantly swims the seven seas, spawning all over the place, and is hard to fish out. 'Nobody has yet managed to dent the population of skipjack. Indeed, I don't think anyone really knows what it is,' Dr Kirkwood told me. That, of course, doesn't mean no one will dent it some day.

The trouble is, skipjack tend to run in shoals with the less fecund bigeye and yellowfin tunas. Yellowfin grows much larger than the skipjack, up to about 2 metres (6½ ft). It provides better-quality, firmer flesh used in sushi and steaks, and is also frozen and canned. It is fully exploited, except in the eastern Indian Ocean, though there are no obvious conservation problems just yet. Then there is bigeye, a deep-water tuna that feeds in the cold depths and matures more slowly than the other tropical tunas, but faster than the cold-water bluefins. It is prized for sushi, and the most valuable species after the bluefin. Adult bigeye tuna run much deeper than yellowfin or skipjack, at around 150 metres (500 ft) down. But juvenile bigeye and yellowfin tend to run with skipjack much closer to the surface.

There is now concern among several management bodies that bigeye is being overfished in large numbers in the Pacific, Atlantic and Indian Oceans. The Red List, compiled by IUCN, lists bigeye as 'vulnerable', which it says means 'considered to be facing a high risk of extinction in the wild'. Dr Kirkwood says, 'Both bigeye and yellowfin need to be managed,' but he knows better than anyone that in the Indian Ocean this isn't happening. The scientific committee of the Indian Ocean Tuna Commission (IOTC), of which he is a member, recommends 'a reduction in catches of bigeye from all gears [i.e. purse-seiners, long-liners and bait boats] . . . be started as soon as possible'.

The concern arises from the way most purse-seiners catch tuna – by setting nets around naturally floating objects or man-made FADs, a much easier way of catching tuna than setting on fast-swimming schools. A sleek, modern boat from either the French or the Spanish purse-seine fleet picks up a signal from one of its FADs that has found the thermocline (where hot water meets cooler water) and begun to congregate tuna. It steams to intercept. It drops a fast skiff, which drops off the net which duly encircles the shoal of fish around the FAD, and the mother ship begins to haul. Alternatively, the purse-seiner may set its nets on whales – ironic, since the Indian Ocean is meant to be a whale sanctuary under international law – which usually escape by breaking through the nets, but sometimes don't. Setting on FADs tends to be highly indiscriminate, Dr Kirkwood told me, in terms of the size of tuna killed and the number of other fish species caught.

Which brings us back to dolphins. What is all this concern about dolphins? They do not seem to be a problem anywhere but in the eastern Pacific, where they run with yellowfins, or maybe it's the other way around. There was indeed a massive by-catch of dolphins in the eastern Pacific in the 1980s, where mostly Mexican tuna fishermen tended to set their nets on pods of dolphins. In 1989 alone, some 100,000 dolphins were killed as by-catch in the eastern Pacific tuna fishery. This problem has, to a very large extent, been solved, thanks to the Earth Island Institute (EII), an environmental group almost wholly focused on the impact of fishing on marine mammals.

The EII's dolphin-friendly programme is monitored and controlled by inspectors and forbids setting on dolphins. But, as a result, fishermen use FADs, which produce up to 50 times the by-catch of 'other fish', including potentially vulnerable species, such as sharks, as well as endangered ocean-going turtles. As a result, other environmental groups, such as Greenpeace and WWF, no longer support the dolphin-friendly scheme and endorse a much cleaner fishing method, monitored by an observer programme, which sets on dolphins but allows them to escape. A conservation-minded tuna expert with a long time in the business told me: 'Thanks to EII, a lot of fishing is done on FADs, which means that instead of dolphins being killed, 20 different species are now in bad shape. They're catching fish that haven't had time to reproduce. From a disaster for dolphins, we have progressed to a whole ecosystem disaster.'

'Dolphin-safe' remains the PR defence of the tuna industry, even in oceans where it has never been a problem. In other words, a lot of credit is claimed by Heinz, one of the market leaders in the canning business, for not doing things in some oceans that have never been done. I called Heinz in the UK to ask whether it was concerned by the massive by-catch of other fish in fisheries using FADs, such as were used by its suppliers in the Indian Ocean. A spokesman replied that the company will only accept fish caught by pole and line and purse-seine (which is what is causing the problem). They do not accept tuna caught by high seas drift-net fishing, pair-trawling or setting on dolphins. This seemed to be missing the point. Drift-netting still goes on, I'm told, but nets over 2.5 km (1½ miles) are now banned under international law. Pair-trawling and sets on dolphins never happened, to my knowledge, in the Indian Ocean.

Whatever its intention, the result of this strategy has been to confuse the consumer about what was really going on by raising animal welfare concerns that were never really relevant, and all the while failing to deal with genuine concerns about the fishing itself. As I have said before, it is just the latest example of species-ism in which animal welfare groups appear not to see fish as animals at all, except as a commodity to be caught in an efficient way. As a strategy for defending Heinz's product against justifiable accusations of destroying the ecosystem and stripping poor countries of fish, such as dolphin fish (not dolphins) and mahi-mahi, that they might otherwise catch, it now seems incredibly out of date.

The only reason there is not more of an outcry about what goes on in places such as the Indian Ocean is that nobody knows about it. Very little information exists about what is caught as by-catch or thrown away as discards in the Indian Ocean. The reason is that hardly anyone has ever collected any. Why? Because it has never been in anyone's interest to do so.

If we look at what happens in the eastern Pacific, where there is an organisation that does a proper job of organising observer programmes and compiling data on what is caught, the figures show that sets on floating objects killed some 237 tons of sharks and rays and 15,500 tons of other fish in order to catch 15,721 tons of tuna, a by-catch rate of over 50 per cent. But, hey, they didn't kill any dolphins.

Sets on dolphins, however, killed only 18 dolphins in seven years, with an additional by-catch of 34 tons of sharks and 295 tons of other fish. Compared with sets on dolphins, from which the vast majority of dolphins are now released, sets on floating objects caused 52 times the mortality. To put it another way, sets on floating objects killed on average 2.6 million other fish and 42,325 sharks and rays a year. Sets on moving schools of tuna came somewhere in between sets on floating objects and sets on dolphins.

So in the Indian Ocean, where there has never been a problem of accidentally catching dolphins, what *have* they been catching accidentally? I spoke to Juan Moròn, head of the Spanish tuna-catchers organisation in Madrid. He said that purse-seining was a relatively 'clean' operation compared to long-lining. So, I said, show me the proof. He told me that some of the Spanish fleet had recently begun to carry observers, at the EU's insistence, as boats in the eastern Pacific have had to for the past 15 years. Where was the data compiled by these observers from the Spanish Oceanographic Institute, which is notoriously secretive about its information? Señor Moròn said he'd try to get it for me. He came back and said the data had not been published.

I, however, knew what to look for; I was just interested to see if anyone was interested in volunteering it. I tried the same question on Dr David Ardill, secretary to the Indian Ocean Tuna Commission, based in the Seychelles. There were some figures headed 'discards' on his website which were said to be 'available from the commission'. So could I see them, please? Nothing was forthcoming, except an email that explained there was little information about by-catch or discards, and what did exist was confidential. 'There are enormous commercial interests involved, and the reality is that if we are not able to guarantee confidentiality, we would simply get falsified data, and would be much worse off than at present,' he wrote.

Surely he could let me see the information that was advertised as 'available from the commission'? Apparently not. I wrote back that I thought his confidentiality rules – which, by the way, are to protect vessels that have found it more convenient (for tax purposes) to fly flags of convenience than the flags of their own countries – looked more like those of a pirates' club than an open and accountable public organisation charged with making one of the world's great commodity foods sustainable. If he wanted people to go on buying cans of tuna from the sea he was supposedly 'managing', he was going to have to do better than this.

Then, to his credit, Dr Ardill thought better of his rules and did start to provide information. But the first piece was not that instructive. It was the observer reports from the voyage of a small Australian long-line fleet targeting swordfish, not tuna, in the western Indian Ocean. Dr Ardill warned me to use 'extreme caution' in drawing any conclusions from the data, which was 'unrepresentative'. Looking at it, he was probably right, but it did show an absolutely huge by-catch of sharks. Those who eat fresh tuna caught on long lines, which can stretch up to 80 miles and have thousands of baited hooks, take note.

All this time the big purse-seine vessels, monitoring their hundreds of FADs and sonar buoys floating in the Madagascan current, and scouring the sea with their long-range sonar, seemed to be getting away. Evidently, no one was going to invite me on board a Spanish or French purse-seiner to count the by-catch. However, I did finally interview someone who had been on such a trip. My contact was Spanish and worked for an internationally known company that made specialist sonars for use in the tuna industry, one of the specialist companies' most lucrative markets.

My contact showed me photographs of FADs – square wooden frames with trailing tendrils of mesh hanging from them – being built by lean, fit men in dazzling sunshine on the deck of a 100-metre (330-ft) super-seiner. This looked less like a fishing boat than a billionaire's yacht. My contact explained that the skipjack swam in the first 50 metres (165 ft) or so of water, the yellowfin down to 75 metres (250 ft), then below that, at 150 metres (500 ft), the threatened bigeye, still shallow enough to be caught in the vast encroaching circle of the purse-seine net, 1,800 metres (6,000 ft) in circumference and 240 metres (800 ft) deep.

The tuna vessel would be monitoring the location of its FADs by satellite phone. Crew would be scanning the horizon with binoculars or checking the bird radar. When the satellite buoy found the themocline, the FADs would be swarming with tuna. Sometimes these FADs might be crude constructions of four sticks and a couple of pieces of wire, and there would be 250 tons of tuna down there, he said. In some circumstances, there could be 40 vessels hunting the same fish and only one would get them. Hence the critical advantage conferred by the best long-range sonar.

In the haul we were looking at on screen, whales were the FAD. Whales, my contact explained, not entirely believably, usually broke the net. It was OK to break the net because of the value of the tuna one still caught. The business was 'enormously profitable'. This haul alone was 80 tons – worth €190,000 (£127,000) – enough to buy a new sonar. But what gave me a thrill of discovery, as I had been looking for it so long, was what was on deck – the by-catch. Gasping their last, as the crew pushed the 90-cm (3-ft) skipjack tuna down a chute to be frozen, were two 4.5-metre (15-ft) manta rays.

It is not just Gordon Ramsay who has a thing about manta rays. Jacques Cousteau, inventor of the aqualung, featured them regularly in the credit sequences of his 1960s and 1970s' television series *The Undersea World of Jacques Cousteau* to symbolise the mysteries of the deep. Divers get excited about these black, wraith-like plankton-eaters with their 'billowing fins'. Like barndoor skate, mantas reproduce very slowly, giving birth to just one very large pup every two or three years. Not enough is known about them to say if they are endangered, but the Red List says 'populations will rapidly decline unless fisheries are carefully managed'.

Manta rays and sharks were a common by-catch, said my purse-seining contact. But the biggest problem was the capture of small tuna of all three main species. These were so small that they were of no great commercial value. What sort of proportion of the catch? 'They never say what they put over the side.' This was the value, he pointed out, of the very latest echo-sounders, which were able to measure the size of the fish before you caught them. The only flaw in that argument is that no one says you *should* use one.

So far we have established that the vessels catching tuna for cans in the Indian Ocean regularly catch at least one vulnerable, large, slow-growing animal, the manta ray, and that they frequently kill bigeye tuna, which is recognised as being on the slide towards extinction. I can now reveal something else, on the best possible authority, which is that juvenile bigeye tuna – a fish as endangered as the Amazon river dolphin, the basking shark or the North Sea cod – end up being canned along with yellowfin and skipjack. A contact at the IOTC told me that this was 'standard commercial practice worldwide'. A tuna buyer at Prince's (owned by Mitsubishi and in my experience much more candid than many others) openly confirmed this. Perhaps that accounts for the reluctance of some supermarkets to say what kind of tuna is in the can. So much for Sainsbury's assertion that there was no question about the sustainability of canned tuna.

We are still at the tip of a large sea mount as far as the number of living creatures killed by the purse-seine fishery is concerned. I am indebted to a friend for sending me the only observer reports ever published on the by-catch of the purse-seine fleet in the Indian Ocean. Ironically, this came from observers with the Soviet purse-seine fleet. The paper, by E. V. Romanov, was an eye-opener. Romanov estimated the catches of tuna in the whole western Indian Ocean at 215,000–285,000 tons between 1990 and 1995 – less than it is now. In the process of catching this, purse-seiners also caught up to the following amounts: 2,300 tons of pelagic sharks,

1,700 tons of rainbow runners, 1,650 tons of dolphin fish, 1,200 tons of triggerfish, 270 tons of wahoo, 200 tons of billfishes, 130 tons of mobula and manta rays, 80 tons of mackerel skad, 25 tons of barracudas, 160 tons of miscellaneous fish and an unspecified number of endangered turtles and whales. Quite a high number of sets were on whales. Many recorded were on whale sharks, the world's largest fish, which have been heavily protected since 2002. Although the Russians fished slightly differently and in different places, the figures are perhaps the best indication ever compiled of what purse-seine fishing for tuna, setting on floating objects, really does to the marine ecosystem.

Romanov concluded that the first step towards solving this problem was for the IOTC to set up a system of scientific control of both the tuna purse-seine and long-line fisheries by means of a network of commission observers on board fishing vessels. Ironically, the Russian purse-seine fleet is still fishing, but is regarded as illegal by the IOTC, which does not allow it to use members' ports.

I sent this study, which dates back to 1998, to Heinz in the UK, asking them whether they still believed the tuna in their cans was caught sustainably. I got an email back after ten days, saying all landings were monitored and all the boat owners were involved in working with IOTC scientists to improve fishing gear in order to limit by-catch. No details on what the by-catch was, of course and no proper answer to the question, which means, sadly, that you and I are none the wiser about the sustainability of the tuna in a Heinz can.

It was only then that I received, thanks to an email, the most comprehensive and eye-opening answer to my question, what gets killed alongside the skipjack tuna that finds itself in your tin. It was an unpublished paper by two scientists from the year 1998–1999 when the three European organisations of frozen tuna producers declared moratoria on fishing in certain areas close to the African coast because of the numbers of (endangered) juvenile bigeye tuna they had been catching. This was the first I had heard of these moratoria. The authors of the paper note that the moratorium months chosen did not have much effect as they did not actually coincide with the *maximum intensity of fishing in the Somalia area.*

The first thing to remark on was that the European fleet was again up to its old tricks and fishing in the waters of one the poorest countries in the world, which currently has no government, and therefore was poorly equipped to defend all its territory. The second thing that stood out, as the Spanish fleet, carrying 40 observers, fished their way across the mid-Indian ocean, fishing on free schools of tuna, FADs and sea mounts, was that some 20 per cent of the catches were of the endangered bigeye tuna.

The observers also noted the scientific names, in Latin, of the species caught as by-catch. I looked up the common names with growing disbelief: it amounted to almost the entire cast list of *Finding Nemo.* Let's start with the turtles, as these are the most endangered. There was a full house of the ocean-going turtles, loggerhead, green sea turtle, leatherback, hawksbill and gulf ridley. These range in their entries from critically endangered (extremely high risk of extinction) to vulnerable (high risk of extinction) in the IUCN Red List. Then there were the whales (remember the Indian Ocean is supposed to be a whale sanctuary): minke, humpback and one the observer could not identify (a pygmy blue is possible in those waters but it could have been anything). Then we come to the fish. There was quite a list, topped by the great white shark, which is now officially recognised on the Red List as Vulnerable. Nothing else was on the IUCN red list in a critical category but many, such as the mobulas and mantas, were listed, showing there was scientific concern about their populations but nobody had yet compiled enough information to list them as endangered. What they did say about these species, the sharks and rays, was that for most of them their resilience to overfishing was extremely low. They were species which reproduced slowly, having a handful of young a year. This applied to the mantas, the mobulas, the stingray, the spotted eagle ray, the shortfin mako, great hammerhead, scalloped hammerhead, smooth hammerhead sharks, the pelagic and bigeye thresher sharks.

Oh, I nearly forgot, they didn't catch any dolphins.

How long the species on the tuna fisheries' by-catch list actually have before they become extinct is hard to say, because no one seems to know they are being wiped out, or in what numbers. But there is plenty of reason to be concerned. Believe it or not, concern about tuna fishing is not just ecological. Though the tuna fisheries operate out in the middle of the ocean, in the last place on earth where you might think there were limits, and target chiefly the skipjack, a fish so numerous that no one has yet counted it properly, they are still no exceptions to Michael Graham's Great Law of Fishing, that fisheries that are unlimited become unprofitable. Entry to the tuna-catching business is still unrestricted. It is still considered anybody's right to buy a vessel and go fishing, and it is a right enshrined in international law. Fleets have grown, and most vessels are operating below their economic optimum capacity. The market was the first to deliver a warning about the world's tuna fleet, both purse-seine and long-line, being far too big in 1999. Catches in that year expanded to 4 million tons, and the market collapsed due to oversupply. The catch returned to around 3.7 million tons in 2000 and 2001, but the catch remains unlimited and may rise if there is more demand. Demand goes on rising. Tragedy is inevitable.

A dire warning about the likely future of tuna stocks was delivered by a senior tuna scientist, in a paper for the UN Food and Agriculture Organisation. Dr James Joseph pointed out that the human population was likely to grow to 10 billion people by the middle of the 21st century, placing even greater demand on the world's natural resources. If there was going to be tuna for them to eat, there was an urgent need to find ways of recording, measuring and limiting the size of the tuna fleet. 'The current legal and political basis ensuring the right of every person to fish on the high seas must be re-examined and brought in line with current reality,' he wrote.

There was a need to limit access to common fish resources, and this might be best achieved by giving, or selling, fishermen property rights over fish in parts of the ocean. Currently none of the regional tuna organisations has the legal authority to allocate or sell such rights (nor, one might add, to police their seas against vessels that are not signatories to their treaties). What was needed, said Dr Joseph, were 'bold new approaches'. Time was limited and action should be swift.

So coming back to the tuna on one's plate, whether caught on a line or in a net, there is a problem. Very little action has been taken to restrict the capacity of the fleet, and some of the authorities have not even attempted to manage the stock and hardly anyone is doing anything about the by-catch of endangered species. Companies, such as Heinz and the big Spanish canneries that supply the European market, appear to be investing little in conservation and to have an ingrained culture of secrecy about by-catch.

The European and American consumer can have limited influence over the Asian long-line fleets, and the fast-increasing bait-boats and illegal long-liners. But they can influence the purse-seine fleets through the big canneries in the Indian, Pacific and Atlantic oceans. Informed consumers could try eating fresh fish that has not had its valuable omega-3 oils taken out, or eat different fish until the canners come up with some better answers about what they are going to do about FAD fishing and the by-catch in the purse-seine fishery generally. At the moment, their policy appears to be: we can't be blamed for what we don't know is being destroyed.

Well, it is possible they do know, and aren't telling, and if they really don't know, it's time they found out.

We consumers will have to consult our own consciences and take what action we can (see page 291). After this, I for one will never be able look at a tin of tuna in quite the same way again. This leaves the chefs and cookery writers to think about the way forward. How are they going to face up to the number of manta rays with billowing fins, the slow-reproducing great white sharks, the whales and endangered turtles that are killed to produce the tuna we're eating? Or to the fact that some of the tuna we buy are themselves endangered? I look forward to finding out.

CHAPTER 12
THE PROBLEM OF EXTINCTION

So long and thanks for all the fish
                                    Douglas Adams

UNTIL A MEETING at London Zoo in 1994, no one had worked out whether a commercial fish species such as the Atlantic cod might be as endangered as, say, the wandering albatross. The realisation by many people outside the cosy little world of fisheries 'management' of what a mess the world's oceans were in dated from that moment. A lot of terrestrial biologists working for the IUCN had read about the Canadian cod crash and felt that they ought to see how fish rated against other endangered animals. Dr Georgina Mace of the Zoological Society of London remembers: 'IUCN had spent four or five years developing listing criteria and somebody said, "This won't work in the oceans". We thought, wouldn't it be nice to get the main people together to tell us *why* it wouldn't work.' She didn't realise at the time the scale of the row she was getting into.

London Zoo had funding for a meeting of 30 scientists from relevant disciplines in species survival to compare the data for fish populations against IUCN's established criteria of extinction risk. 'Two weeks beforehand, the word got round all the fish management people.' The word 'tuna' was on the list, so the Japanese said they wanted to be there. Then Spain, Portugal and Australia said they wanted to send representatives. The fishing nations had realised that a process was going on outside their control. Dr Mace remembers: 'The Japanese were just too persistent. They wore me down. They came under a whole set of conditions; it was a scientific meeting, they had to bring new data. They didn't protest at the meeting, but the moment they got home they began to attack the thing. [Fishing] is a huge global industry and they were not about to let it fall apart.'

A lot of people were unhappy about the idea of listing wide-ranging species with a high reproductive rate, such as cod, as in danger of extinction. Among them were marine biologists who looked at larger, slower-reproducing fish, such as sharks, which had previously been the only focus of concern. Some scientists said that it was theoretically possible for a cod laying 10 million eggs to repopulate an ocean. Dr Mace was resolute. She said, 'Any species where the number of adults is going down, must be in trouble. Do you have criteria which pick that up when it's too late or when things are getting worse?' The IUCN criteria say that if a species has declined by more than 20 per cent over ten years, however big its population, it is 'vulnerable' to further depletion. Cod and haddock had declined by more than 20 per cent over ten years, as had bigeye tuna, so all three went on the Red List, alongside the cheetah, the wandering albatross and large-leafed mahogany. Elodie Hudson, who worked with Dr Mace, went to a world fisheries conference in Brisbane. 'It got rather hideous,' she remembered. 'We went through the list over a day, with lots of shouting.'

Canada spent a lot of money hosting the IUCN General Assembly in 1996. The Canadian Department of Fisheries and Oceans (DFO), found out that the 1996 Red List was going to list Atlantic cod. IUCN was asked to withdraw its list. Georgina Mace was forced to meet someone from the DFO, who said he was outraged that the whole assessment had been agreed without the scientists who 'managed' the cod. The Canadians, for whom everything to do with fish is political, asked how an assessment could possibly be done without calling in all shades of scientific opinion. 'We said it was very simple,' said Dr Mace. 'We just do it.'

Then the Japanese got up in the first session and protested about tuna. She remembers that there was a lot of bad behaviour all round, but some fishing countries were 'duplicitous in the extreme'. She reflected after it was over: 'The species that caused the most controversy were based on UN Food and Agriculture Administration [FAO] data. You can dispute the risk of extinction, not the data.'

There was indeed a fair argument to be had over whether the IUCN's criteria exaggerated the extinction risk for commercial cod and haddock, or for other fish stocks of which millions of individuals still survived in the oceans. There is no recorded case of a marine fish species going extinct – as there is of large sea mammals, such as Steller's sea cow. But, then, how would you know the last fish of a species has died out? There is something known as commercial extinction, not a very exact term, meaning that a fish has declined to such a level that it is not worth fishing for. Yet that term is hardly useful for species such as bluefin tuna, where the price goes up the fewer fish there are in the sea. Fish such as cod are distributed over a large area, so true extinction may not be that likely. Or is it? Canada's cod has proved the existence of the Allee effect (under-population) in fish. There are terrestrial species that are like cod in population terms, such as locusts and some plants. The North American passenger pigeon was the most plentiful bird species on Earth, yet was gone in five years. The American bison was nearly wiped out by hunting. Elodie Hudson says, 'You can't assume that extinction is any less likely in the sea.' That hasn't stopped scientists from big fishing countries stomping around trying to develop endangerment criteria that exclude fisheries.

The first encounter in what promises to be a long war between two sides who each think they know better, fisheries people and wildlife people, was a win for the furry animal brigade. (Given their growing, and resented, interest in fish, perhaps we should now call them the scaly animal brigade.) The big bang of the cod crash of 1992, which made few front pages in Europe at the time, has rumbled on in the mind of the public, in the pages of *Science* and

Nature, and into disciplines far removed from fishing. People as different as economists and television producers have began to realise that oceans are interesting and that fish are not just scaly things that don't appeal to animal welfare groups, but beautiful and fascinating creatures. Political scientists have realised that conserving fish is one of the world's most challenging intellectual problems. (Every discipline has a habit of screwing up over fish – at least once. Witness the missed opportunity to do fish issues justice, while showing stunning pictures of the deep, by the makers of the BBC series *The Blue Planet.*)

Slowly, a lot of people are asking themselves the big question: if palatable fish are not going to be an infinitely renewable resource, what is everybody going to eat when the world's population reaches 10 billion? Come to that, if the fish are going to run out, isn't that a concrete example of a limit to human growth – as the authors of the FAO report warned in the 1970s? And everyone said they were wrong? So maybe they were right in their Malthusian assumption that human society would need to limit itself or face mass starvation. Or will fish farming and technology make up the gap? Those are things we will discuss, but first let's try to simplify things even further.

It shouldn't be that difficult to run a fishery responsibly and sustainably. Surely it's easier than splitting the atom, landing a rocket on the moon, fixing the hole in the ozone layer, managing a currency, or decommissioning the world's nuclear arsenals. You would think so. But if the failure to do it is anything to go by, it's harder than any of these tasks. How do I know? Because in the course of writing this book I have flown around the planet and surfed the net looking for a single country where they have done it without making terrible mistakes that crashed some of their stocks, perhaps irrevocably, and I have been mostly unsuccessful. Even in the countries where they are doing OK, there are huge problems, question marks about the future, and failures to think about the survival of *all* living creatures in the sea. So how can you say anyone is yet managing the ocean sustainably?

The problem is relatively simple. People have got used to using space technology and weapons of mass destruction in our marine ecosystems but we haven't developed workable ways of limiting them, though we pretend we have. We haven't even worked out what is down there that we wish to protect, let alone what our grandchildren would resent us for not saving. What makes it hard to manage the oceans is that fishing, the most destructive thing going on in them, is nearly everywhere the responsibility of the most junior politician in the cabinet – the fisheries minister. In some countries he's not even in the cabinet. Everywhere in the world the fisheries minister is there just to perform the traditional role of keeping the fishing industry happy. If you've read this far, you would probably agree that this approach isn't working. The question is what to do next.

Here it must all get personal, for otherwise it would get too complicated. I am a journalist by trade and therefore carry in my mental toolkit the opinion that experts may at all times be talking through the back of their necks. But you have to start somewhere. It is foolhardy to ignore the wisdom of people who have spent a lifetime thinking about a problem, even if they have made mistakes. They might, conceivably, have learnt something. So I will begin by examining the wisdom of the organisation that ought to have been fixing the world's fisheries problem, but sadly has not managed it yet, the UN Food and Agriculture Organisation in Rome.

The FAO has been rightly derided for believing duff information from the Chinese and failing to blow the whistle when the world's fisheries went into decline. But it continues to collect mostly good-quality information and does much good work. It comes up with codes of conduct for responsible fishing, and plans for dealing with illegal, unregulated and unreported fishing, which nation states do too little about. In its shamelessly pro-development way, the FAO tells the truth, though not necessarily all the truth or all the time.

Think about the FAO's current warning that 75 per cent of the world's fisheries are fully exploited, over-exploited or significantly depleted. That is about as strong as a pro-fisheries organisation is going to get to saying that the world's fish stocks are going down the plughole forever unless we do something about it. You may wish to think hard about its statement that the western central Pacific and the eastern Indian Ocean have the only under-exploited fish resources on the planet. Surely what the FAO means is that the fish populations of those oceans are relatively healthy compared with every other ocean? So should a responsible UN body really be suggesting that more fisheries should be developed where there are fewer rules than anywhere else on Earth? As long as you remember that the FAO betrays a verging-on-insane bias in favour of commercial fisheries, you may read on.

The FAO decided to consider whether there were common trends in fisheries that were becoming unsustainable and, if these trends could be found, whether there was anything one could do about them. So it invited some of the world's most eminent fisheries scientists and economists to Bangkok in February 2002. At least the food must have been nice. The centrepiece of the discussion was a paper by Stephen Cunningham and Jean-Jacques Maguire that hit a most un-FAO-like note of alarm. It said that the FAO's current assessment of the world's stocks – that 75 per cent were fully exploited, over-exploited or significantly depleted – was likely to be conservative. 'Considering the number of [fish] resources on which little or no information is available . . . the situation may be even worse than it appears.'

If you wade through the conclusions reached by these fine minds, you will find analysis characteristic of the FAO at its best. First, all the experts present were agreed that the factors contributing to unsustainability – that ugly but necessary word – in fishing were similar in almost all jurisdictions. They identified six types of pressures:

Inappropriate incentives – subsidies and other economic incentives.
High demand for limited resources – reflected in the rising price of fish as a luxury, health-conscious, fashionable food.
Poverty and lack of alternatives – if, like penniless fishermen in the South Java Sea, you have nothing and no job, you will carry on

fishing long after anyone would have paid you to so that you can eat. This is the hardest problem of all.

Complexity and inadequate knowledge – by this the experts meant, I think, both the lack of adequate scientific data on complex fisheries, such as the West African up-wellings, and human legal systems that are too complicated to be understood properly.

Lack of governance – a good old UN euphemism, this, for corruption, fiddling, incompetence and failure to enforce the rules.

Interactions of the fishery sector with other sectors and the environment – interaction is a characteristic FAO euphemism for buggering things up. Fishing sure has.

The gurus of hunter-gathering also came up with a list of new ideas intended to address the world's problems. A few of these are platitudinous, but not many. I'll give their headings, then translate, as before.

Rights – the FAO's worthies were of Garrett Hardin's view that we should fence the commons by granting rights, or property rights. They called for the allocation of 'secure rights to resource users (individually or collectively)' for use of a portion of the catch or a portion of the space taken up by the fishery. This is a vast change, the implications of which are discussed in Chapter 13.

Transparent, participatory management – this doesn't happen in the EU or many other places right now. Some countries are well ahead already. Iceland, for example, gets its fishermen to do an annual scientific trawl survey, which means they are more disposed to believe its conclusions.

Support to science, planning and enforcement – slightly platitudinous, this.

Benefit distribution – now this is an interesting one. It means that fishermen should pay 'rent' for using the fishery to the people who own it, i.e. the public, possibly on top of the income tax they pay. This works only if the fishery is profitable, which only those run on economically efficient, though not socially inclusive, lines are. This is an important suggestion that would compensate, to some extent, for the downside of handing out rights to the fishery, which, over time, would concentrate quota in just a few hands.

Integrated policy – I think this means that each nation should set explicit objectives that address all aspects of sustainability, such as the conservation of marine wildlife as well as fish. This would mean the creation of some marine reserves. Why not say so?

Precautionary approach – this, I suspect, is UN diplomatic code for saying that none of the world's fisheries institutions should be managing stocks for Maximum Sustainable Yield any more, as many tuna bodies are still doing. They and many other fisheries should be setting total allowable catches, according to FAO guidance, so that fish stocks would recover even if the worst climatic or population fluctuations previously recorded were to hit the stocks.

Capacity building and public awareness raising – again, necessary but platitudinous.

Market incentives – by this they meant they agreed with eco-labelling schemes for the consumer that identify sustainable fish against unsustainable fish. Not a bad idea at all.

The assembled gurus also had some good ideas about governance. They made a strong plea for the creation of institutions to manage the high seas where these did not already exist. Where they did exist, they said these institutions should be strengthened considerably to play their necessary role. Each should receive, *as a minimum*, clear legal authority to manage and conserve fish stocks, and both the political mandate and the resources to ensure that fishermen complied with conservation policies and regulations. They meant that the day-to-day requirements of fishery managers should not be politically negotiable, as they are in so many countries. Thinking about their recommendations, I can think of few countries in the rich north, let alone the rich south, that have institutions of this kind today, least of all the EU. So this is quite some recommendation.

As a way of speedily immersing oneself in the latest ideas about saving the world's fish, this summary is hard to beat. But I must add two observations from my own journey around the world: the problems are worse than many people imagine and the solutions are more controversial. It is one thing, for example, to recommend the use of rights to control fisheries, another to persuade people to accept them over time and make them work – as we will find out in the next chapter.

You miss the gossip, the sense of what the key points were, by merely reading conference conclusions. So I spoke to Jake Rice, one of the participants. Jake, remember, is a Canadian. This means he signs up to fishing being a cultural (i.e. subsidised) activity, as well as an economic one. He goes misty-eyed about the guys trying to make a living in all those outports in Newfoundland – though you and I know they are subsidised up to the hilt.

From something Jake said, I suspected there was a real conclusion they all came to, over the Thai food and beers, that was too difficult for the FAO to write up. Jake admitted this. He said you probably wouldn't find a bald and heartless sentence saying so in the conclusions because the FAO, being a UN body, had to have 'some sense of hope in its texts' – whatever that means. This, however, is what they really concluded in Bangkok: 'You have to be willing to write off one of three dimensions – ecological, economic or social – to solve the problem of sustainable fisheries.'

Let's examine that proposition because this is as close as we are going to get in fisheries to a solution. This is what I think Jake and his colleagues meant. There are three options:

A. You can have a fishery that is healthy ecologically and economically, but you have to forget about supporting fishing as a cultural (i.e. subsidised) activity all the way round the Scottish, Iberian or Newfoundland coast.
B. You can have a fishery that works economically and socially but not ecologically (but presumably not for very long because the resource will be gone in a few years once you have mined it).
C. You can have a fishery that works ecologically and socially but not economically (conceivably by making its money out of something else, such as tourism or nature conservation grants, like paid graziers on a nature reserve).

Only A and C work because B is not sustainable in the long term. My view is that A may well have a future in offshore waters, while

C may have more of a future in inshore ones. I think what the FAO guys really decided, between the Thai food and the evening drinks, is actually quite profound. It is an answer to the meaning of life in a world of constantly expanding technology.

It may be profound but it is not wholly original. Jake's workshop aphorism rang bells. It reminded me of something John Pope, then a senior scientist at Lowestoft and subsequently an adviser to the Canadians on their cod stocks, told me in 1997. 'You can't maximise the economic and the social and the ecological all the same time.' In other words, something's got to go.

So what's got to go, you may ask, if we are to have fish forever? I think the answer, almost certainly, will be different in different places. But in most industrial fisheries more than a day's journey from the shore, I'm beginning to think I know what's got to go because it's gone everywhere else in raw materials industries the world over. And that's people.

Why should fishing be any different from mining, shipbuilding or farming, especially if it's destroying the world that belongs to the rest of us that we do not want destroyed? What is special about outports that means they should survive? The bottom line is that there are just too many fishermen, and fishing technology gets better every year.

Jake Rice believes in the cultural side of fishing, and therefore thinks there is something we want to hold on to in coastal communities. He thinks there is something about those outports that needs to be preserved *as it is*. He may be right. But I am not so sure I want to preserve the Scottish, Iberian or Newfoundland kleptocracies one little bit. So I will let Jake, who is a more compassionate soul, put the argument himself, for he put it beautifully, and far better than I could, in an email he sent me:

> I grew up in an area of dairy farms, with small towns scattered every few miles apart – each providing the feed mill, milk plant and general store for the neighbouring farms. Now agribusiness owns more than 90 per cent of the farms, and the little towns are either empty or gentrified by people who

commute an hour each way to the city to work and shop. A rich, interesting culture was lost, and no one blinked. Other examples abound. It is a genuine mystery why coastal fishing communities have such a uniquely sacred identity, but the pattern repeats in Canada, the USA, the UK, Denmark, Iberia and the rest of Europe at least. I really don't know why – maybe better musicians and songwriters.

Now we're getting somewhere, I thought. My view, as the son of a man who was both a farmer and a miller, whose family's businesses are no more, is that we live in a different world. There is plenty of employment, if you adapt, without having to hoe cabbages by hand, which I did in my youth. You can't go back. Jake, who contributed to the crash of Newfoundland cod, but acquired some wisdom in the process, is an incorrigible old sentimentalist. He loves fishermen, just as the world used to. But can we afford to be sentimental about fishermen if the price of having too many of them is the destruction of fish and the wonders of the oceans, which we are only just beginning to understand?

We need a new approach. People only make a virtue of necessity if you stop subsidising them to do the wrong thing. In Lowestoft, where there is no fleet and very few fish any more, they produced The Darkness, a retro glam rock band currently taking the world by storm, and tourism is taking up the slack. We used to think miners should be subsidised. We know farmers *are* subsidised, but we realise it is a disgraceful subsidy. Very soon we will think the same about subsidising fishing.

But that's enough theory. I really think it's time for a reality check, to see if these ideas work somewhere out there where they are trying to save the cod.

CHAPTER 13

DEATH OF THE COWBOY

REYKJAVIK, LATE SEPTEMBER. Flecks of the first snow of autumn drift out of a leaden sky above the Hvalfjord, the whale fjord. Here on the capital's waterfront one of Iceland's big freezer trawlers, the *Therney*, lies offloading her frozen, filleted cargo of redfish and cod in cardboard boxes labelled in Icelandic, English, Spanish and Japanese. Fishing forms a significant part of the economy in Iceland, and, as a result, the stock market takes the annual fish-stock assessments very seriously, dipping if the cod assessment is down. Kristján Loftsson, a part owner of the *Therney*, who has brought me to the dock in his huge Jeep, a characteristic vehicle for a well-to-do Icelander, points to the front page of Fiskifréttir, Iceland's equivalent of *Fishing News*. It shows an Icelandic trawler chugging back to port, its hold so full of blue whiting that the boat sits low in the water and looks like a submarine. Kristján shakes his head. The picture annoys him because he sees it as a stain on Iceland's reputation. Icelanders think they know as much as anyone about conserving fish.

They are almost certainly right. Icelandic waters are among the few places in the world where cod stocks are both relatively plentiful and, arguably, on the increase, though the upturn is in its infancy. After several years of setting quotas low, and enduring criticism for doing so during a general election, the government has just agreed to increase the quota by 30,000 tons, in line with scientists' assessment of what stocks can stand and their conclusion that big year-classes are beginning to come through. This increase in allowable catches is equivalent to two-thirds of the entire spawning stock of cod in the North Sea. Here in Iceland it is a kind of banker's Christmas bonus, bringing the total allowable cod catch up to 209,000 tons, and making those few people who own quota to catch it even richer than before. It will be a windfall of £52.2 million for the Icelandic economy, a tidy sum among a population of only 300,000 people. The injection of cash could push prices higher than they are already, and move Icelanders further up the list of the richest people in the world.

Iceland's ability to manage its fish stocks starts with its isolation. In fact, its distinctiveness hits you when you get in the shower and smell the sulphur in the water from the hot springs. But Iceland's ability to determine unilaterally what goes on for 320 km (200 miles) around its coast does not necessarily make for good management. The years from the end of the last cod war in 1976 – when this country without a navy expelled the British and German trawlers that had overfished its stocks of cod – to 1984 are now just a bad memory. It was after the last cod war that overfishing by Iceland's own freezer trawlers, newly built on British lines, replaced overfishing by foreigners, rather as happened in Newfoundland after the declaration of their 320-km (200-mile) limit. Iceland took a stronger grip of the situation than Canada, proving perhaps that countries for which fish represent a sizeable proportion of their economic interests take more focused action than countries with more numerous ways of making a living and only a social interest in fishing. 'In Iceland we know that if we mismanage fisheries, we are out,' says Hjörtur Gislason, who writes about the fishing industry in *Morgunbladid*, the leading Icelandic newspaper, and was the first European journalist into Newfoundland when cod fishing was banned in 1992.

It hasn't been easy. Iceland's Marine Research Institute publishes a 'Blue Book' every year, showing the state of every stock in Iceland's territory, from whales to whelks, via the real revenue earners, such as cod and haddock. The Blue Book is a great exercise in transparency and public information. If you look in it, you will find that cod, Iceland's most important stock, has been much more plentiful than it is now. In the mid-1950s the fishable stock was estimated to be over 2 million tons, and the spawning stock over 1 million tons. There were some nasty shocks as a result of overfishing from 1980 to 1983 and in the early 1990s. Then, around the late 1990s, scientists drastically overestimated the spawning stock, a mistake they realised and admitted in time. Iceland cut back its cod quota from 300,000 tons to 150,000 tons, which caused grumbling, but the fishing industry tightened its belt and put up with it in anticipation of better times ahead. For three years, one during a general election in which the ruling coalition was re-elected, the resolve held firm at the ministry and the institute. Now the cod numbers appear to be rising again and one hopes Iceland has not been premature in awarding fishermen more quota.

There are some striking contrasts with other parts of Europe. In Iceland, where the fishermen are mostly rather rich, they accept the science, rather as major public companies dealing with pharmaceuticals would, instead of perpetually grumbling, as their poorer European brethren do, that there are more fish in the sea than the scientists see in their trawl surveys. Does this tell us something? Who honestly challenges the science – rather than the interpretation of the science – in any other discipline and is taken seriously? Would you have an operation by a consultant who didn't take the science seriously? Would you buy a car built by someone who didn't? Jóhann Sigurjónsson, director general of Iceland's Marine Research Institute, says: 'Iceland is perhaps the only European country that is still based on hunting. We are heavily dependant on the resource. So Icelanders have accepted in principle that science-based management is the only way to go. We may argue about it, but at the end of the day there is a consensus. We have general support for our recommendations.'

One reason for Icelandic fishermen respecting the scientific process of counting fish is that they themselves are involved in surveying the stocks. For a few days a year each trawler is supplied with nets of the kind they used in the mid-1980s, which are mothballed somewhere in Reykjavik, and these are used in the annual trawl survey, alongside results from grids trawled by the institute's three impressively equipped research vessels. Some people grumble all the same that changes in the climate mean that the fish are no longer where they were in the mid-1980s. They may well have a point. Last summer was so hot that the scallop population collapsed. Monkfish, a species usually caught only on the warmer south coast, has been caught all round the island. Juvenile haddock, saithe, plaice, mackerel and cod have all thrived since 1995, when the sea became warmer, but there are also fears that climate change may be interfering with the breeding success of cod in Icelandic waters, and may be the reason why the population is not increasing faster. There are many unknowns, even in one of the best-studied fisheries.

You do wonder whether Dr Sigurjónsson and Iceland's fisheries minister, Árni Mathiesen, should not have left the quotas lower if their aim to let the stock recover and grow to former levels is more than rhetoric. Crucial to this growth is to build up the numbers of all-important older spawning cod, the 'old ladies' as Dr Sigurjónsson calls them. He is surprisingly downbeat when I talk to him, and I suspect a political accommodation may have been forced on him. He says: 'The stock is healthy. Some people would be very happy. We are not happy unless we are achieving 100 per cent of what we want. We are advising now not on a crisis but on the best level we want the fishing to hit.' Am I being too cynical, or am I noticing that it is at precisely this point that conservation, even in the best fisheries in the world, runs out of steam because there are so few parallels for holding on tight and restoring the stock to its former abundance?

As Odd Nakken, a wise Norwegian fisheries scientist and former director of the research institute in Bergen, put it to me seven years ago: 'The history of conservation that hurts is only 20 years old.'

Dr Sigurjónsson will tell you that fishermen are still catching the cod too young. The majority of those caught are six years old, and cod only begin to spawn at five around Iceland. He would rather be catching the country's quota from cod aged seven, eight or even nine years old. A nine-year-old cod can weigh 5 kg (11 lb), so the number killed would be fewer for the same weight of fish. Nature equipped the cod to live to over 20 years old and to more than 90 kg (200 lb) in weight to spawn for longer. The old ladies will lay their eggs for a longer period – up to six weeks. These eggs, which are healthier and stronger than those laid by smaller fish, stand a better chance of hitting the optimum water temperatures for survival. For the young cod to thrive there must be plenty of food. Dr Sigurjónsson has plans for closed areas to protect some of these old ladies. He says the cod recovery that we are now seeing has been achieved by sacrificing the prawns, which are worth more than the cod itself.

I am still surprised to see just how fine the line is between responsible fishing and overfishing, even in a system where politicians and scientists have the mandate to make tough decisions. And how hard these decisions are, given growing climatic variability and a principal stock, cod, that is perhaps a quarter of its biomass 50 years ago. The recovery of cod, haddock, saithe and plaice is more tentative, the variability just as great and sustainability less assured, even in one of the best-run fisheries in the world.

Worries aside, the Icelandic system has more precaution, more meaningful penalties and more flexibility to close areas where juveniles are caught than most others. This extra caution is written into the law in exemplary fashion. Since 1995, the total allowable catch of cod has been limited by law to no more than 25 per cent of each year's spawning stock. Compare this with nearly 100 per cent of adults (and quite a few juveniles) being killed in the North Sea in the 1990s and you are not surprised that there is barely anything left in those waters. Another admirable precautionary measure built in to the Icelandic system is that an annual increase in total allowable catch of more than 30,000 tons is not permitted, just in case the scientists have got it wrong.

The Icelandic system has a lot of other things going for it, too. The basis of the tremendous economic success of Iceland's fishing industry since 1984 was the introduction at that time of a revolutionary, rights-based system of quotas, which give fishermen a long-term investment in the health of fish stocks. Just what the FAO worthies now recommend as the best way of running a fishery. Iceland's system of individual transferable quotas (ITQs) has been as successful as it has been controversial. (Don't get too bothered by the name, of which there are variants around the world. It just means you own that quota for good unless you sell it.) As with any new system, there have been winners and losers. The problem that some people perceive in this one is that it has tended to concentrate quota in the hands of large companies. People who have played the system skilfully have made large sums of money out of the rise in the value of quota. A handful left the country and now trade quota on their mobile phones from Florida or Guernsey while doing no fishing at all. Meanwhile, coastal communities that were slow to buy enough quota for their needs, or were already in debt, ended up having to sell their trawler quota. (In the inshore waters there is, quite sensibly, a different regime but one based, controversially, on effort controls – limits on the time that can be spent at sea.) The system in place remains extremely popular with those who operate the big freezer trawlers beyond 20 km (12 miles) of the shore. They will tell you that it has, quite simply, changed the way that fishermen think.

Iceland's tradeable quotas are not absolute property rights, as were allocated in New Zealand. They amount to a right to take a certain proportion of the stock, depending on scientific advice, for the foreseeable future. This, fishermen say, has brought an end to the race to catch fish before somebody else catches it. Fishermen, therefore, are more likely to leave fish in the sea if they think prices are poor at the moment, for they are sure it will be to their advantage in the long run. As Árni Mathiesen put it to me: 'You have somebody who thinks of the quota as their property, who thinks it has to be as good tomorrow as it was yesterday. That is what matters in the long run.'

Fridrik Arngrímsson, chief executive of the Federation of Icelandic Fishing Vessel Owners, told me: 'Since ITQs were introduced, the thinking has been completely different. We don't think of volume so much any more, we think of value. Once a skipper caught 80 tons and showed it to me proudly. What he did not say was that he was getting nothing for the cod because it was all twisted and bruised.' Fishermen who fish for the maximum economic efficiency fish less, burn less fuel and drag up the bottom less as part of the bargain.

Acquiring the right to a certain amount of fish, come good weather or bad, has changed not only the way fishermen think, but the way they fish. Rúnar Stefánsson, the manager of the Grandi fleet, which owns the *Therney*, tells us that fishing is no longer a rush to catch the most fish, but a matter of catching high-quality fish and selling them for the best prices.

All 16 of Iceland's major commercial stocks, from plaice to mackerel, are now managed under ITQs. People in Europe who say the Icelandic system wouldn't work in a mixed fishery like the North Sea – or along the coast of West Africa – don't understand what they are talking about. The rules are strictly enforced by the directorate of fisheries, a government body, and the coast guard. Indeed, one of the major advantages of the ITQ system is that there is an incentive for fishermen to watch their neighbours to ensure that others are complying with the rules, for they will lose out if the fish stock were to decline and the value of the quota were to fall. There are 'black' landings, but these are thought to be far less than they are in the EU, perhaps under 3 per cent of the cod catch.

Stringent rules and enforcement are crucial to the system. The rules say that everything caught must be landed (no throwing away juvenile fish below the size limit, as in the EU). Discarding is banned. Regulations on mesh size are supposed to mean that juveniles are not caught. All the same, Iceland's waters are a mixed fishery, so fishermen are not always going to catch what they set their nets for. The system therefore allows some flexibility. If you do not have quota for what you catch, you may buy it to make your catch legal. Or you may transfer some of your cod quota to catch, for example, monkfish. You may not use monkfish quota, however, to catch more sought-after cod. If nobody at all wants to sell you quota for your over-quota cod, you are allowed to land your fish, and 80 per cent of the value of the over-quota catch will go to the Marine Research Institute. The system is backed up with tougher penalties than in other regimes. Skippers who break the rules run a real risk of not being allowed to fish, with serious economic consequences. Licences can be withdrawn temporarily or permanently. Skippers convicted of serious offences go to jail.

Supporters of the ITQ system are evangelical about the way it has improved the psychology of fishing and transformed fisheries investment. 'The whole world is heading this way,' says Ragnar Árnason, economics professor at the University of Iceland in Reykjavik. 'If people have property rights, they look after them.' What appear to be the most successful, i.e. the most stable and sustainable fisheries in the world – hoki in New Zealand, rock lobster in Australia and black sablefish, halibut and (up to a point) pollock in Alaska – are now run on this basis. Professor Árnason is probably right.

Over dinner in Reykjavik's Hotel Holt, with Ragnar Árnason and my friend Orri Vigfússon, probably the greatest campaigner for the Atlantic salmon, I ask whether we can go over it again. Ragnar says the people who own the rights will defend and protect the system. But is a right to catch a fish the same thing as owning a piece of land? Fishermen in Iceland still pay mortgages and are governed by what economists call the discount rate, i.e. it is better to have money in the bank tomorrow than in a year's time. Although I am increasingly convinced by what Ragnar says – that ownership is the only system that gives a value to the fish left in the sea – I decide to play devil's advocate. I ask whether there are not just as many successful examples of running a fishery around the world (few though these are, viewed overall) that are based not on rights but on the principle of a managed common? Orri indulges me while I come up with three examples. I decide to rate them, and Iceland's system, out of ten.

The North Sea herring
The history of the North Sea herring in the 1960s and 1970s was

reminiscent of the gold rush on blue whiting today. The traditional way the British and Dutch fished for herring, with a drift net, came to an end soon after the Second World War. The bottom trawl, targeted at spawning concentrations of herring, was the instrument of destruction in an international free-for-all. Norwegian, Icelandic, Bulgarian, Polish and East German vessels raked the sea beyond the 10-km (6-mile) limit. Britain joined the Common Market in 1973, placing the whole North Sea under the management of the Common Market and Norway for the first time, but it was still impossible to reach agreement among all the people who fished for herring.

The stock collapsed in 1977 and the fishery was closed. The next year the much smaller west of Scotland herring fishery closed too. The fishery reopened in 1981 after a partial recovery. For the first time it was governed by a system of total allowable catches. Unfortunately, the closure from 1977 to 1981 virtually destroyed the market for herring in Britain – people had got used to eating other fish – but not in eastern Europe. So vast Russian factory ships, known as 'klondikers', with fleets of catching vessels, fishing on licences bought from the European Community, still fished the herring, as did the Scottish, Danish and Norwegian pelagic fleets.

In 1996, the herring was again under pressure. A 50 per cent cut in quota was agreed and enforced because the memory of the crash was still alive in people's memories. It seems to have worked. The pelagic sector is one of the EU's few successes. Its fleet comprises a few, highly equipped boats and appears relatively under control. Stocks are healthy, though not the great bounty they once were.

Marks out of ten: *five* – just for the herring, not the whole EU Common Fisheries Policy/kleptocracy.

Barents Sea cod

The cod stocks shared by Norway and Russia in the Barents Sea are the largest left in the world. In the Lofoten Islands, stepping stones that stretch out into the Atlantic just inside the Arctic Circle, you see racks of dried cod or 'stockfish' drying in the breeze in the same way they did a thousand years ago. The Barents Sea's huge resource of cod sustained the whole of Europe through trade with the medieval Hanseatic League, but it is now somewhat depleted. The total quota agreed by Norway and Russia in 2003 was about 440,000 tons, but one should remember that the Barents Sea used to produce catches of 900,000 tons a year in the 1970s. Lately stocks are down, but not as low as they were in the mid-1980s, when catastrophe was only narrowly averted. In 1986 and 1987, thanks to the overfishing of capelin for fish-meal for Norway's salmon farms, the cod suffered a food crisis and cannibalised their young. Contrary to scientific predictions, the cod stock did not grow. In a decade of bullish estimates it declined rapidly. By 1989 the cod was in bad shape, and scientists, led by Odd Nakken, recommended a drastic cut in the quotas. There was resistance from fishermen, but a determined fishing minister, Oddrun Pettersen, from Finmark, the most fish-dependent county in Norway, banned fishing for capelin and slashed the cod quota to 200,000 tons, shared between Norway and Russia. For the first time Norway threw subsidies at fishermen not to fish. It paid the mortgages on the fishermen's vessels.

By 1996, when I visited the Lofoten Islands, there was plenty of cod again. But Odd Nakken was unhappy when I met him for a drink in Bergen. Norway operates a system of total allowable catch and quotas, not unlike the EU's. Norway and Russia's ministers had agreed a total allowable catch of 850,000 tons of cod for that year. Too high, observed Nakken, puffing his pipe. About 100,000 tons too high. So it proved. Norway's subsidies had mothballed the fleet, which simply got going again. Paying fishermen not to fish, without reducing the size of the over-large fleet, merely postponed rather than cured the problem. The cod stock has tumbled since then to a worrying extent. I am told that Norwegian fishermen suspect that the Russians are cheating on their total allowable catches and quotas, so they are avoiding their own fairly stringent rules. Norway is also exporting its subsidised fishing capacity by catching blue whiting in the current free-for-all. Its share of the carnage last year was 700,000 tons. Many voices in Norway say it would be better off with ITQs. The government does not want them because it says this would be inconsistent with its policy of supporting coastal communities in the north.

Marks out of ten over the decade: nine at the beginning, but subtract six for going wrong since: result *three*.

The Faroe Islands cod

The Faroese cod stock is healthy – just look at the faces of Scottish fishermen when boxes of fat Faroese cod are landed at the Scottish port of Scrabster beside the puny specimens from the North Sea. Like Iceland's, it appears to be increasing for climatic reasons. Faroese fishermen traditionally catch cod by long-lining or 'jigging' – a technique that involves jerking a silvery lure up and down on the end of a line – rather than trawling. Jigging is now done electronically by several machines attached to the side of a fast inshore boat, making it a light, efficient killing technique. Faroese cod has produced catches of around 30,000–40,000 tons a year, reasonably reliably, for years – except in the disastrous years of 1993–8, when catches plummeted to around 6,000 tons for reasons that have never been satisfactorily explained, then bounced straight back up again. Incidentally, until that time 20 per cent of the entire government budget was spent on fishing subsidies. Since then, these have been stripped out.

The Faroe Islands operate a system of effort control, under which each vessel is allocated a number of days at sea. Like Iceland, it also uses closed areas to protect spawning aggregations of cod. The quota is divided up and a committee works out how many days each vessel can fish. Interestingly, the Faroese had a system of ITQs imposed upon them by Denmark after their cod crash. ITQs did not have the backing of the fishermen, so they did not work. After a re-think, controls on effort – time spent fishing – were the industry's choice and now these seem to work but economists, such as Ragnar Árnason, justifiably say the system is less economically efficient. A fishing vessel could find itself sitting in port for half the year. The system restricts a certain size and gear type to particular areas around the Faroes. All fish caught during the trip can be landed and legally sold. There is the usual problem that there is always some aspect of fishing that effort controls do not constrain – whether it is the sonar, the kind of gear, or the power of the vessel. And it is human nature for fishermen to increase these. Interestingly, the Faroese parliament has introduced a way of compensating for 'technological creep' – the annual improvement of gear, boats and sonar. It lowers the number of fishing days by 2 per cent a year.

The only thing obviously wrong with the Faroese system is the amount of cod it awards itself – 33 per cent a year of the total biomass of the cod, which scientists from ICES and the Faroes' own fisheries lab say is too high. The Faroese parliament prefers the advice of its own scientists. ICES says that the Faroes ought to build a more precautionary approach into its stock assessments. While cod stocks continue to rise, apparently for climatic reasons, they can stand the pressure, but the Faroes could find themselves in trouble very swiftly if those climatic factors were to change.

Marks out of ten: *six.*

The Atlantic salmon

As well as these three examples of how fisheries have been managed with some success in a common sea, I can't resist adding the example of Orri Vigfússon's buy-out of the Greenland, Faroese, North East England and now some Irish drift nets for Atlantic salmon. This example cuts across categories – common sea fisheries and rights-based fisheries, recreational and commercial forms of fishing – but solutions in conservation, I have noticed, invariably come from lateral thinking. The decline of the Atlantic salmon in the 1980s set anglers thinking about the overwhelming number of threats to migratory fish – habitat degradation in rivers, poaching, estuary nets, and, most intractable of all, high seas netting on the salmon's annual migration to feed off Greenland.

Orri, who owns Iceland's vodka factory, was the only person to do something about the largest and most intractable problem, single-mindedly setting up a body called the North Atlantic Salmon Fund and calling on the owners of river fishing all over Europe to protect the value of their properties by buying out commercial fishing capacity. The Inuit fishermen of Greenland were not making much money out of drift netting for salmon, which was stopping several hundred tons of salmon returning to European and North American rivers where they would have made considerable revenue for the owners of sport fishing rights. The Greenland fishermen were happy to be handed money and grants, through their fishing association, to re-organise themselves to tap more profitable fisheries.

A deal with the differently organised Faroese fishermen followed and a deal with 52 out of the 68 licensees of drift nets off the North East coast of England (worth £3 million to the fishermen) followed that in 2003. Orri's river in Iceland has seen salmon numbers soar following the Faroese net buy-out. The River Tweed, on the border of Scotland and England, happened to have its best year since the 1960s after the North East drift-nets were bought out. The key to all these deals seems to have been that Orri and his supporters negotiated with the private interests that exist even in a common. In some cases, like the English North East fishery, the law actually had to be changed to give the commercial fishermen secure rights before these could be bought out.

Marks out of ten: *nine.*

How does Iceland compare?
Before Ragnar, Orri and I attempted to compare how Iceland rates in comparison with the Faroes and Norway, we discussed the downside of the Icelandic system, which I had been discussing with opposition politicians. Because of the way quota allocation is decided, the acceptance of this system within Icelandic politics is more fragile than outsiders imagine. Kolbrún Halldórsdóttir, an MP of the small Left-Green party, put the problem succinctly:

When the quotas were given out, they were a gift. Then they got a value. Big companies bought the quotas from small companies, and the small villages had no fish coming into their harbours. That became a reason for people moving away from the villages and into small towns. It angers us to see people who are filthy rich and living in Guernsey putting their money into shopping malls in Reykjavik instead of into the fishing industry. Sustainable development is not about collecting money into one big pile. You have to reduce the ability of the company to access it in big piles. You have to recognise that the plan does not work and recognise the imbalance it has created in our society.

The Left-Green party have come up with a thoughtful way of restructuring the system of ITQs over 20 years so that the quotas are tied to communities rather than companies. This has been received as ingenious by all shades of opinion. But, crucially, *it is still based on rights.*

Other opposition parties look to the Faroes for solutions. Magnús Thór Hafsteinsson, an MP for the small Liberal party, told me: 'There is a human factor in all this. The small communities that depend on fish are losing the battle.' Gudjón Kristjánsson, a Liberal MP and former skipper, says: 'The winners are always winning and the small ones always losing. The fisherman can sell his quota, but who will buy it? The big company.'

Hafsteinsson points out that the system of tradeable quotas was meant to lead to a contraction in the fleet, whereas the decrease in the number of large trawlers has been disappointing. This is a reasonable point. In fact, the number of licensed vessels has declined by almost half since 1984 and the fleet tonnage has also declined, but the decline in the fleet has been much less than you would expect from the reduction of fishing effort within Icelandic waters. The reason is simple: many of Iceland's fleet of big, super-efficient 70-metre (230-ft) trawlers now fish both in Icelandic waters and for redfish, blue whiting and other species in effectively unrestricted fisheries beyond the 320-km (200-mile) limit. Iceland has turned from a coastal waters fishing nation to a buccaneering force upon the high seas, not yet like Spain, but a different kind of fishing nation from the one it was. The Liberals would like to move to effort control, with the effort divided increasingly between the smaller vessels.

This is a popular message, and one that is actually happening by default in Iceland, but it threatens to undermine its entire ITQ system. Back in 1983, when transferable quotas were handed out, the politicians of the time bought peace with the owners of hundreds of small inshore vessels around the coast by allowing them free fishing. They originally took only 3 per cent of the catch, but, strangely enough, that percentage grew and grew. A series of rules restricting the size of inshore vessels proved counter-productive. It prompted a technological revolution. Fishermen sold larger boats and bought faster, smaller ones equipped with the very latest in powerful engines, electronic cod jigs and other gizmos. The coastal boats now take more than 25 per cent of the quota. They always catch more than they are supposed to, but 25 per cent is still not enough for the genial giant Arthúr Bogason, who represents the inshore fishermen.

Charming and apparently 'green', Arthúr wants more, and, over a delicious lunch at his expense in one of Reykjavik's finest hotels, very nearly convinced me that he should be given it. Unfortunately, it is the job of writers to bite the hand that feeds them, and I must report that if there is one part of the Icelandic system that is out of control, it is Arthúr's. The inshore fishermen have become greedy, and they are undermining the sustainability of the system. Although the inshore boat is capable of producing top-quality fresh fish, which is then whisked to the shore within the day, a system of effort controls means that there is an inbuilt temptation to catch as much cod as you can and rush it home without icing it properly.

Iceland is a crucible of competing theories for controlling fishing, which is why I went there. Which system is the best? I think on the evidence you have to conclude that the system of tradeable rights has the edge over effort controls because it gives a value to fish that are left in the sea. If rights work, they give fishermen a financial interest in conservation. They also give the fishermen who hold quotas an interest in reporting other fishermen who break the rules, and in fishing in a different, more careful way. In political terms, having an electorate of owners rather than supplicants takes part of the responsibility for achieving conservation away from politicians, and vests it in companies, which is no bad thing, as politicians have failed everywhere in their duty to conserve fish. Rights-based systems take the responsibility for decommissioning away from treasuries – which never want to fund this anyway – and give it to the market. It is easy to think of circumstances in which a rights-based system might not work, such as if fishermen resented the way it was imposed, as they did in the Faroes. Rights are not a panacea for all the ills in fisheries. On the other hand, a private company in New Zealand recently went to the fisheries minister and asked for a *reduction* in quota. Where else would that happen but in a rights-based system?

Effort controls, by contrast, give an incentive to fish as hard as possible in the time allowed. And they encourage technological creep. In Ragnar Árnason's words, 'Effort reduction does not work because fishing effort has so many variables. You can always find the dimension of effort that is not constrained and not constrainable.' The greedy, hi-tech inshore fleet – which surfs on a wave of nostalgia in Iceland for a time when fishing was done by small entrepreneurs, not major businesses – is a problem Iceland must address before it undermines the whole system. I leave it to the journalist Hjörtur Gíslason to defend the status quo: 'I think you are stuck with the system. The little boats can't go out too far or in bad weather. They are controlled by "days at sea", so they fish as much as they can. When you do that you take too much each day and the quality of the fish is not good. The trouble is there is not enough fish for everybody. It does not matter which system you have. That is limited.'

Marks out of ten to Iceland's ITQ system: *eight* because I still have anxieties about the way the allocation of quota was done and the failure to reduce the size of the fleet. I give the inshore, effort-controlled sector a generous *four* because they are catching too much.

So much for Iceland, but what about the rest of the world? Do ITQs have to be as socially divisive as they have been in Iceland, I ask Ragnar Árnason. It's not great advertising for a system of property rights at sea if one sees a few fat cats, in Hjörtur Gíslason's phrase, 'frying their bellies in Florida' in 20 years' time. Professor Árnason remains relaxed about the inequalities of wealth that have come from the share-out of quota in 1984: 'It is a fundamental point of human nature that you can't create new wealth without a quarrel over how to share in those benefits. The ITQ system increases the economy overall, so many people will benefit. It is the most efficient way biologically and economically of using the resource. It does not solve poverty, but then what industry does?' Once the quotas are profitable, they can be taxed, like any other area of the economy, and the revenue used for programmes to help the poor and dispossessed. Iceland is introducing 'rents', i.e. taxing ITQ holders, in 2004.

That didn't quite tackle the problem I foresaw, so I asked the question a different way. Could he really live with the kind of inequalities that the ITQ system seems to engender in somewhere like Senegal, where the alternative, for non-fishers, is starvation? Árnason replied:

You can solve all equity problems at the outset. You can choose who to give the quota to, who is eligible. The ITQ system is neutral. You can achieve any desired distribution of the benefits through the initial allocation of quota rights. If there is a perception of inequality, that should not be blamed on the system, but on its implementation.

He says you can give the quota to a cooperative or a beach – albeit with an inevitable loss of economic efficiency. If that is not feasible for some reason, you can tax the companies who do own the quota to provide a social security system for the poor, or to provide training so they can get more skilled jobs. You can only expect so much from the fishery, and the poor are always with us.

In Alaska, where the halibut fishery was once run under effort controls, the year's quota was sometimes fished out in 24 hours. Now the halibut fishery is run under a rights-based system, which allows fishing to go on at a low level all year. It is much more profitable, produces better-quality halibut, kills fewer fish in the process and gets round the 'belly-frying' problem very neatly. The quota is attached to the boat owner, and he has to be on the boat when it is fishing. He may fry his belly if he likes, but he won't get that warm for much of the year in the Bering Sea or the Gulf of Alaska. There is also a rigorous restriction on the vessels that are licensed, which makes it more difficult for there to be technological creep.

Despite the demonstrable advantages to conservation of giving fishermen rights to fish stocks, there is profound suspicion of rights-based solutions among environmental groups. Ironically enough, in the USA, the citadel of capitalism, there is a huge defence of common ownership going on, by people who seem unversed in the total failure of common ownership almost everywhere. Even the distinguished Daniel Pauly, who sits on the board of at least two environmental groups, is opposed to selling off rights to fish because, he says, they are public assets that already have an owner – the public. He favours instead an auction of rights every year to the highest bidder. I can't see for the life of me why that wouldn't lead to the same race you get with effort controls to catch as much fish as possible in the time available – but it's the only point on which we disagree.

I tend to agree more with Rashid Sumaila, a fisheries economist in Daniel Pauly's own unit at the University of British Columbia. He does not see why the ownership of the sea, or its assets, should not be managed as private property any more than the land. For ownership of land confers a peculiar blend of rights and responsibilities under most jurisdictions. Certainly under English law, one may own land, but one has a responsibility for it. An owner may not pollute his land and may be subject to all sorts of environmental laws and designations as he goes about his business. Under English law, one may not prohibit the public from enjoying a right of way over it; indeed, it is a responsibility that comes with the privilege of ownership that one should keep up any stiles, trim any hedges that inhibit the right of way and safeguard any areas listed for their unusual species. An owner of land may be taxed, on behalf of the public, to provide rent for the public treasury.

Sumaila is a pragmatist. The test, he says, is whether rights would improve the fishery. In Norway the problem is one of policing and trust and proper enforcement. Ownership of rights, by giving fishermen an interest in enforcement, would improve the fishery. In Senegal, he says, social scientists would allege that even if you allocated quotas initially to everyone with a *pirogue*, people would sell their quotas too low because of poverty. The benefits are less clear, the dangers more obvious, but the problem might go away if the quotas were owned by cooperatives.

Whether environmentalists like it or not, the idea of rights to fish seems to be sweeping the world. If they deliver benefits for fishermen and for fish conservation, which they seem to, I think we should be in favour of them. John Gruver, a former skipper with the Alaskan pollock fleet, now working for United Catcher Boats, a small boats' association based in Seattle, drew a parallel between the US prairies and the North Pacific: 'In the 1880s they invented barbed wire and divided up the ranges. Now everybody accepts that's the way it is. Nobody likes to see the death of the cowboy. Nobody will think it should be any other way with the sea in 20 years' time.'

So far I hope I have shown that the problems in the world's fisheries are more extensive than is often imagined. Most of the management solutions are controversial and complicated, and will always be vulnerable to an over-generous assessment of the stocks. There is one solution that is controversial with fishermen, particularly with those who think they own rights to fish, but it is simple and totally effective. That is preventing fish from being exposed to any kind of fishing gear at all.

CHAPTER 14

DON'T FEED THE FISH

GOAT ISLAND MARINE Reserve, Leigh, New Zealand. The first bank holiday of spring. My half-sister Gillian, a New Zealander for more than 30 years, collected me off an overnight flight from Los Angeles and drove me the 90 km (55 miles) north of Auckland to Leigh. As we began the hour and a half drive, the Pacific morning air was chilly but clear. The sun was bright and increasingly warm as we negotiated the green hills, the indented coastline and its mangrove-lined estuaries – the first of many scenic surprises for any European traveller who expects all temperate zones to look the same. Suddenly, as we followed the prominent signs to Goat Island Reserve off the highway and drove towards the sea, I could see why an adviser to the fisheries minister had assured me that the southern hemisphere's equivalent of Easter Monday bank holiday was actually the best time to visit the reserve. The place was crawling with people.

Around a car park on a grassy hillside there were people hiring out wetsuits and flippers of all sizes. In front of Goat Island, a tree-covered hillock 200 metres (180 yards) offshore, were families of all ages swimming or wandering about on the sandy beach and rocky terraces. Parties of teenage trainee divers arrived with their instructors as the water began to warm up. Groups of tourists were embarking in a glass-bottomed boat, the most comfortable way of seeing the explosion of fish and underwater vegetation that has taken place since the reserve was created in 1975. At the end of the track, in front of the buildings of Auckland University Marine Laboratory, two figures were waiting for us: Thelma Wilson, regional director for the department of conservation, a suntanned, uniformed outdoor type in shorts, and Dr Bill Ballantine, godfather to the world's marine reserves, a wiry man in a green marine lab sweatshirt, smoking one of many roll-ups.

Leigh is perhaps the world's best example of what natural ecosystems look like when they are left alone, without fishing pressure, for a long time. The reserve has exceeded its founders' original expectations, and not just from an ecological point of view. Before it was earmarked as a reserve, Goat Island was a popular fishing spot for fishermen. Conflict arose with the marine lab because, as Thelma puts it, 'people kept eating the experiments'. The reserve was created, after 12 years of lobbying by Ballantine and his colleagues at the university marine lab, for narrowly scientific reasons. Ballantine is the first to admit that he never foresaw what an attraction the reserve would become. As we chat in the sun, he points to some of the younger visitors among the 500 or so present that day. 'During all the years we were seeking arguments for doing this, I never heard anyone say schoolkids,' he grinned. 'The reserve came first and the people came after – nobody expected them.'

We stand looking at these unexpected beneficiaries of science. Most are families who have driven out of Auckland for the day to stand in the clear sea water and gawp, as a profusion of fish nibble and swim round them. Others have come to snorkel, hiring a wetsuit on the beach for a few dollars, or to stare through the glass panel of the boat at the fearless shoals of large snapper, blue maomao and spotties that swim above forests of kelp, only yards from the shore. It is the only place I know of where you will find signs saying: 'Please do not feed the fish.' These were introduced by Thelma in response to an unexpected problem. What do people do when they find unaccustomed numbers of wildlife in a picnic spot? They offer them a sandwich, of course. The signs explain that feeding can make fish sick, can teach them to bite and otherwise modify their behaviour. There is plenty of natural food for them in the ocean.

What happened to the ecosystem was also unexpected. Snapper are the most prized sporting fish on this coast – and for that reason increasingly small and scarce. In the 547 hectares (1,370 acres) of the reserve, the largest snapper are eight times the size of the snapper outside. They are also 14 times more numerous. Brochures tell you that you will also find butterfly perch, silver drummer, porae, red moki, leather jacket, blue cod, red cod, goat fish, hiwihiwi, butterfish, marble fish, red-banded perch and demoiselles – all swimming around without fear of people and within a few yards of the shore. Indeed, they have so little fear that they may nibble you to see if you are food.

Most of this marine menagerie is readily visible to an inexperienced 11-year-old snorkeller in a hired wetsuit – popular with parents because a child dressed in one tends to float. Further out, in deeper water accessible to more adventurous divers, are delicate gorgonian fans, lace corals, sponges, sea squirts and anemones. Under the kelp forests, hidden in holes and crevices in the rock ledges, are big rock lobster, or crayfish, much larger than those in the commercially fished waters outside. A line of orange floats stretching out to sea, easily mistaken for the boundary of the reserve, turns out to belong to lobster pots legally set there by local fishermen. Although excluded from the reserve itself on pain of a NZ$50,000 (£20,000) fine, fishermen are happy because they catch as many rock lobster on the reserve boundary as they would on many more miles of coastline.

When the reserve was created, explains Ballantine, there was nothing particularly special about the marine ecology. The rocky coastal reefs were known as 'rock barrens' because nothing grew on them. The most common bottom-living species were large sea urchins, which graze on kelp. As in other parts of New Zealand's northeast coast, the kelp forests had virtually disappeared by the 1960s. The connection between the disappearence of the kelp and overfishing only became clear when the Goat Island Reserve had been established for many years. At a certain point, the undisturbed snapper and crayfish reached a size at which they could prey on the *kina*, or large sea urchins, that fed on kelp. So the kelp forests gradually returned, bringing in turn food and shelter for many other species of fish and shellfish. Biologists call this a 'trophic cascade', when the recovery of predators at the top of the food chain has effects that flow down to lower levels. Some of the changes happened quickly. Others took much longer. Some are still happening. It took only three years, in a much larger reserve set up around the Poor Knights Islands to the north, for the snapper population to recover to a level eight times more numerous than before a no-take zone was imposed. The full recovery of kelp forests at Leigh has taken 25 years.

The public's interest in the reserve, almost negligible at first, has grown as the transformation of the marine habitat has become more extraordinary. In ten years Thelma has noticed more family groups coming for the day out. Nobody needs an aquarium when they have a reserve like this. The reserve has changed people's idea of what is interesting in the sea. 'When we started none of the students would ever dive here – it was boring,' said Ballantine. They all wanted to go to Australia to the Great Barrier Reef. Now students are bobbing about all over the place. 'Thelma's one fear,' grinned Ballantine, 'is that a shark swims into the bay and she's got to explain to everyone that it's protected.'

Like many people who have fostered a big idea that is sweeping the world, Ballantine tends to be regarded with indifference or suspicion in his native land. Now his convictions have borne such fruit, he is prone to a certain inflexibility of attitude. He can be a grumpy old bugger, refusing to answer questions if he doesn't like the way they're asked, flaying questioners who don't believe him – recognisable signs, perhaps, of a lifetime spent making a case to people who disagree with him. In his retirement, he is allowing himself the luxury of not being polite to the people he disagrees with. Whether this is wise is another matter.

Now the New Zealand government has introduced a parliamentary bill that would, among other things, allow the setting up of marine nature reserves in a shorter period than the ten years it has taken, on average, to establish each one of those it has so far. 'We may have 18 reserves,' says Ballantine, 'but all *crept* through.' Now the bill is before parliament, and Ballantine is the Aunt Sally of the fishermen again, getting a regular verbal bashing for his opinions and giving as good as he gets in return. I suspect this is the explanation for his take it or leave it attitude. The fishermen fear him, though, because they know his message is popular. To the conservationists, he is a guru.

Creating a network of marine reserves is obviously an important thing for its own sake. But I am determined to get Ballantine to talk about the potential benefits for managing fisheries, as this is key to getting them accepted politically. And how practicable would it be to manage, say, the North Sea using closed areas alone – as Han Lindeboom suggested all those years ago when I first walked into the wrong lecture in The Hague in 1990. But Ballantine's being grumpy and doesn't feel like answering the question, even though he has written papers on the subject. He says it is the wrong question. 'Whether reserves would help fisheries is not the point.' The point, he says, is that there should be areas of the sea where we see what nature is like in the absence of human transformation. The people who talk about sustainable fisheries, he says, do not really know at what level the ecosystem should be sustained. 'One possibility is that you are merely sustaining the wreckage. It could be twice as good, or ten times as good. Without reserves, the check is not possible.' I realise I have been trying to stop him making a really good point.

OK, I say, and decide to go with the flow because Ballantine is on a roll. He is focused now on the prevailing official mindset that fishermen are entitled to hunt everywhere in the sea. 'We're still assuming that you have one management plan for the whole sea. There is no biological, economic or social theory that supports this. At no time did anyone say it is a good idea to let everybody fish everywhere.' He tells me that at a public meeting he rounded on one sport fisherman opposed to reserves with this question: 'Let me get this right: you are saying you want more than 90 per cent of the ocean for your own amusement?' History does not relate if the sport fisherman managed to splutter a reply, but it sounds as if, in tennis terms, Ballantine had served an ace. People talk about fishermen being 'stakeholders' in the sea, says Ballantine, but that ignores the fact that fishermen may be foreclosing all sorts of options for future generations. 'My grandson,' says Ballantine, whose daughter married and divorced a fisherman, 'is the stakeholder.'

We are back to the old question: who owns the sea? Fishermen like to think it is them – especially when they have been given or sold property rights to exploit a sustainable proportion of the stocks, as they have in New Zealand. But the real answer to who owns the sea is everyone and no one: if there is an owner of a common resource in a democracy, it is the people. Citizens have, until now, had few ways of exercising any influence over what happens in the sea. The voice of the citizen is seldom heard over the voices of the 'user groups' – commercial fishermen, sport fishermen, fishing associations and lobbyists – and establishment fisheries scientists, another interest group without, necessarily, the same interests as the public.

What people like Ballantine have done – and it is a very considerable thing – is to circumvent one of the most intractable problems of the oceans, which is that fish don't vote and fishermen do, so politicians go on giving in to fishermen. He has done this by talking to the public over the fishermen's heads, telling them that they own the sea. It is powerful stuff. 'I can go into a room full of businessmen, housewives and boy scouts and say we have to keep some of the sea for itself, and it is only the managers and the scientists who are against it. The public are on side. People say, "We naturally assumed you would be doing that." ' Ballantine has never been more aware of how powerful his message is than at a time when the government is proposing to beef up legislation on reserves – and encountering fierce opposition. 'I am finding myself in the position of telling politicians how to buy votes. You can get popular with a million voters by declaring a reserve, and unpopular with 500,' he says.

Ballantine says he began to notice evidence of a shift in the usual pattern of inertia when public opinion stopped Shell from dumping the Brent Spar, a redundant oil platform, in the North Atlantic in 1995. As he wrote at the time:

> Slowly but steadily the public is starting to say – We don't *have* to chop *all* the trees that could profitably be made into useful things. We don't *have* to mine *all* the land that contains useful minerals. We don't *have* to dump rubbish in *all* the available spaces because it would be cheaper. Very soon they will be saying – We don't *have* to fish *all* the sea just because it would be fun or profitable. They are already doing so in New Zealand and the idea is likely to spread.

By now it already has.

Attitudes change. 'A hundred years ago people did not go into the bush without a dog and a gun, otherwise people wanted to know what you were doing,' says Ballantine. But attitudes have changed more on land than they have so far in the oceans. The advantage of reserves, such as Leigh, is that they enable people to look below the veil of the waves and see fish as wild animals, not as fillets on a slab.

One begins to see why Ballantine's ideas are so revolutionary and why fishermen find his ideas so frightening. No-take zones may not have happened in many places yet, but the idea seems at times to be sweeping the planet – a transformation as big in its way as feminism. Ballantine actually thinks of the idea in this way, likening the change in our view of the ocean to attitudes to votes for women. Why are there numerous marine reserves in New Zealand, yet so few in the mainland USA, and virtually none in Europe? He says: 'Revolutionary ideas get adopted in small ways miles from the centre of established thinking. The first place there were votes for women was in Wyoming territory. One of the last places to give votes to women was Switzerland.' A hundred years on from women getting the vote, students cannot comprehend the arguments used at the time for not giving it to them. 'If you were organising a democracy in which half the human race was excluded today, which half would you exclude?' he asked with a grin.

Ballantine has a point: the way things were organised was not necessarily right, proper or fair – it merely reflected the status quo, which people felt must be right. Women had a hell of a job organising things differently, but they did it. Gradually, attitudes to the oceans are changing, too.

I wasn't going to let Ballantine get away with providing no answer to the question of what benefits reserves have for fisheries. As we were talking, I found what I wanted in literature put out by the New Zealand Department of Conservation. I have described already how snapper on reserves get bigger and more plentiful. Bigger fish produce more offspring. So, of course, do more fish. The snapper egg production on a healthy reserve is therefore vastly greater than in the sea outside. Snapper are estimated to produce 18 times more eggs on reserves than in other parts of the sea. In fact, snapper egg production on 8 km (5 miles) of marine reserve is equivalent to egg production on 145 km (90 miles) of unprotected coastline. Surely this means that there is potential for exporting larvae to supplement and replenish the sea elsewhere, if only enough large reserves were created?

The argument raised against that idea by conventional fisheries scientists – even the conservation-minded Sidney Holt – is that fish move about. You would therefore need reserves that covered vast areas, including entire migration routes, to protect them. In fact, snapper move about too, and the received scientific wisdom said an increase in the snapper population on New Zealand's marine reserves wouldn't happen. People thought the snapper that got big in the reserve would simply go out and get caught. In fact, it doesn't happen like that, although studies carried out at Leigh and elsewhere show that half the population does take part in a seasonal migration.

If reserves work for snapper, you might ask, could you design a reserve for a highly migratory species, such as the bluefin tuna? Ballantine replies that the usefulness of reserves much smaller than a species' range in conserving its general population has been demonstrated for over 100 years on land. The first reserves, established around the 1850s, were for birds. These generally protect only one stopping-off point of a bird's migration, but no one complains that this means they will be ineffective. Scientists Ransom Myers and Boris Worm have found global 'diversity hotspots' – Serengetis of the ocean – where migratory species, such as tuna, turtles, marlin and sharks, congregate. These, they say, would make ideal marine reserves.

So if reserves are a good idea, where should you put them and how big do they need to be? (Ninety per cent of the literature on marine reserves, says Ballantine, is about what they would or wouldn't do for 'target species'.) The fact is, he says, that choosing reserves because you want to protect this or that is a mistake. He says the fundamental thing is to protect some of the marine environment. We know too little about it to know what the effects of protecting it will be, so his view is to just get on with it and set up a few areas, preferably with some close to where people live. 'The idea that we know where to put them is worrying,' he says. I sort of know what he means, but it seems to me to be a lot easier to have a reason. It makes more sense (and it is more accurate and honest) to tell a fisherman, 'We are designating this area because we *believe* on the best possible evidence that it used to be enormously rich and could be again, with many spin-offs for fishermen, if there were no fishing.' Should one listen to the fishermen when designating a reserve? Ballantine says: 'If we want to do the right thing, we just do it. We wouldn't take any notice if it were, for instance, the construction industry on land. We would just do it.' Perhaps fishermen aren't treated just like any other industry often enough.

How much of the sea would you need, then, if you really wanted to conserve fish and used reserves as a deliberate way of boosting the world's fish production? Ballantine doesn't have to think for long. 'For science and education, 10 per cent. For proper conservation of species, 20 per cent. For the general good of fishing 30 per cent, but if the sea were to be intensively used, you need 50 per cent. You need another bay completely no-take for each one that is used.' I am thinking, what sea deserves the description 'intensively used' more than the North Sea? Looking at the tiny area currently devoted to no-take-zones underlines the scale of the task involved in creating more. Scientist Daniel Pauly has published research saying that if you wanted to take the world's fish stocks back to where they were in the 1970s – never mind the last century – you would need 20 per cent of the oceans.

It is arguable that good fisheries management by conventional means is actually unachievable. So many, like Pauly, would argue that reserves, probably quite large reserves, are important not only in themselves but for fishermen. From what we know about fish, these sanctuaries could become repositories of large, super-fertile mother fish, laying large, healthy eggs that can restock the oceans.

There is certainly a lot of emerging evidence (published by WWF) that reserves can improve fishing. The Soufrière Marine Management Area in St Lucia, West Indies, increased local catches by 46 per cent from traps, and by 90 per cent in fishing grounds around four reserves in five years. One of the oldest examples of a reserve that had a tremendous effect in conserving fish is the Merritt Island National Wildlife Refuge in Florida, established in 1962, when it became the security zone for the Kennedy Space Center at Cape Canaveral. After nine years of protection from fishing, the reserve began to contribute world-record-size fish to the surrounding recreational fishery. The evidence of reserves producing fisheries-enhancing effects around New Zealand, however, is equivocal. Leigh is one of the few reserves to produce a measurable spill-over effect with its spiny lobsters. Experimental fishing around Long Island-Kokomohua Marine Reserve found that blue cod had risen there by 300 per cent over seven years, but they remained constant 1.6 km (1 mile) or more from the boundary. It may be that the species is so sedentary that any effects will be felt as the export of larvae. This is extremely difficult to measure.

All the same there is enough evidence that reserves or no-take-zones could improve fisheries to make you think that fishermen would be open-minded about them – right? Wrong.

In New Zealand, where the Labour party's Helen Clark leads one of the greenest governments for some time, the Marine Reserves Bill has its back to the wall on the creation of more reserves. At the Department of Conservation (separate from the Department of the

Environment, which, for some reason, is not that supportive on marine issues), a huddle of officials had stacked together every possible scientific paper for me on why marine reserves were good thing – expecting the same sceptical reception from me as they get at home. In fact, I was already convinced from what I had seen.

Ranged against their bill are, in order of political difficulty, Maori, the fishing industry and recreational fishermen. Maori and the fishing industry object to the reserves essentially for the same reason – that they are big shareholders in commercial fishing operations. I had a tense lunch with Daryl Sykes, who runs the formidable spiny lobster fishermen's association. I happened to remark that it was probably inevitable that there would be a massive expansion of no-take zones all over the world, since this was a commitment signed by 180 or so nations at the Johannesburg summit in 2002. 'You want to create a whole load of marine zoos?' asked Daryl. 'Why?' The temperature of our conversation leapt to combustion point.

As far as Daryl was concerned, this wasn't a done deal at all. He was a big man, in top-to-toe denim and cowboy boots, and he sat alongside me on a bench seat, penning me in and stubbing his fingers at me for emphasis. 'There will need to be compensation for loss of access and fishing opportunity,' he growled. Rock lobster quota now trades at NZ$230,000 (£85,000) a ton. Rock lobsters do live in rocky ledges along the coast in the sort of places that end up being marine reserves, so Daryl had a point, though how good a point was arguable. You could just as well say that the fishermen were given their quota, so why should they be compensated for small reductions in the area they could fish? The quota system, in any case, only confers the right to harvest a sustainable quota within other constraints, including marine reserves. Daryl's ferocity was inexplicable – except perhaps against the backdrop of New Zealand greens, some of whom I had met, and whom my medic sister described as 'over the top'. To me, Daryl, who had charged off to answer his mobile phone, all denim and machismo, was a redneck, though a very smart one. He and the whinging Kiwi eco-freaks deserved each other.

The other person present at our tense lunch was Max Hetherington, who represented the recreational fishermen. Max, a chain-smoker in his sixties who had already had one heart attack, was a determined spear fishermen, despite the evident risks that diving posed to his red-haired frame. He pointed out that recreational fishermen proposed the first three marine reserves in the South Island, but now they feared a proliferation of reserves up sensitive coastline – with some good reason. One reserve has been proposed only 1 km (½ mile) along the coast from an existing one. The fishermen had started to fight. As a sport fisherman myself, in fresh water, I found it difficult to understand how the New Zealand government had managed to fall out with the recreational fishermen, who should be the natural allies of marine reserves. It all boiled down to the fact that the Department of Conservation didn't believe in anything short of a no-take zone, having spent years with partial restrictions on fishing around the Poor Knights Islands, which had little effect.

This kind of absolutism, which may well have come from Ballantine, seemed bound to cause trouble because it was seen as an erosion of ordinary people's rights. I pointed out to Max that what the Department of Conservation was doing was actually contrary to terrestrial practice. You would never create a terrestrial national park that way in Africa or India. There would have to be buffer zones where there could be some sustainable logging, some hunting, some settlement – as there is in the great rainforest national park at Korup in the Cameroon, for example. Buffer zones are now thought to be an essential tool for enlisting the support of local people – but why only in Africa and India and not New Zealand? It seemed to me that a few areas where commercial fishing was banned, but recreational fishing was not, could have bought crucial support for conservation.

*   *   *

Back in the UK, I was reminded just how far ahead New Zealand already is in protecting the sea. One of the places I had long thought there should be a reserve was in Lamlash Bay on the Isle of Arran – the place, you may remember, where the best-known sea angling festival in Scotland was once held. In the 1960s, competitors from all over Britain would expect to catch a total of around seven tons of fish over three days. The festival last took place in 1997, when the total catch was 13 kg (28 lb) of fish. Lamlash Bay is not unlike Goat Island Marine Reserve, having a large hill rising out of the bay, called the Holy Isle, which is now owned by Tibetan monks and was once a retreat for Christian holy men. The area is very picturesque and has always been a stop on Glaswegians' day trips 'doon the watter', when they travel down the Clyde in ancient paddle-steamers. One of these, the *Waverley*, still steams through Lamlash Bay to Campbeltown, where Tommy Finn's fishing operation is based. Arran is still the Glaswegians' holiday island, and a marine reserve would do wonders for the island's economy. The wildlife there, which included shrimps, tiny flounder and cuttlefish in the shallows, eider ducks billing at each other in spring, wheeling gannets diving at mackerel in the summer, and dogfish and cod all year round, always made me think that Lamlash Bay would be the perfect place for a protected area when we spent our summer holidays there with my wife's parents, and I nurtured an ambition to propose it one day.

On my return to England, I received an email saying that someone had actually done so. I was delighted and wrote off in return to the proposer, called Howard Wood, offering any help I could provide. An organisation called the Community of Arran Seabed Trust (COAST) wanted to establish a small no-take zone from which no marine life could be removed, to allow nature to regenerate within the area. They also wanted to declare the rest of the bay as a marine protected area, restricting clam dredging and commercial diving for shellfish. This would be the first no-take zone in Scotland and the first community-proposed one in Britain. In England a very small no-take zone has been set up, by the government agency English Nature, around the long-established marine reserve (where fishing was still allowed) at Lundy Island in the English Channel. The Lundy one was top-down. The Lamlash one would be bottom-up, the result of people power. It germinated out of a meeting of COAST's chairman, Don Macneish, a decade earlier with – you guessed – Bill Ballantine.

The second time I got in touch with Howard Wood things weren't going so well. COAST now had the support of 500 people – a quarter of the population of the island – but they had had a meeting with somebody called Gabriella Pieraccini, the devolved Scottish Executive's head of inshore fisheries. She had said that the campaigners would have to have agreement from the Clyde fishermen before the project could be completed. The Clyde Fishermen's Association was opposed to the idea. So much for the people owning the sea. Howard Wood, a diver, said COAST had received some support from Scottish Natural Heritage, the statutory conservation body, when they had pointed out that part of the sea bed was an old, wrecked maerl (calcified seaweed) bed thousands of years old, a favourite nursery area for fish and shellfish. The maerl bed would recover if it were left alone. The bay itself had long been protected as it was used by the navy for submarine manoeuvres and as a mooring place for larger vessels, but now the shoals of 'queenies' (small scallops) were being fished out fast. Mr Wood remembered that these queenies used to rise up from the sea bed like clouds of butterflies when you passed by. Recently he found a clutch of a dozen together, more than he had seen for years.

Howard Wood said the impression he had been given was that the Scottish Executive – which happens to be at the centre of a complaint from the European Union for not applying the law with sufficient rigour to stop fishermen cheating – was trying its best not to upset the fishermen. Meanwhile, two scallop dredgers had arrived in Whiting Bay, just a short distance along the coast, and had been scraping the bed of the bay for scallops all week within a few metres of the shore. Local people, he said, were outraged that this was legal and that they could do nothing about it. The Clyde Fishermen's Association told them that it was nothing to do with the residents of Arran what happened on the sea bed off their shores.

The idea of leaving parts of the sea alone is very simple. It cuts across the ideas of traditional scientific fisheries 'management', with its impressive-sounding professionals telling us how much they know. Scientific fisheries management has failed almost everywhere, sometimes because of some factor that scientists failed to predict, sometimes because politicians wouldn't listen to them. The beauty of large reserves for biodiversity and fish management is that they are an insurance policy against this kind of failure, and a reminder of how marine ecosystems behave in the absence of human transformation. It is obvious from experience that they work. Monitoring them with satellite technology is a piece of cake. You make fishermen fit transceivers. If you find a fishing boat has been in a reserve without any extenuating circumstances, you take away its licence to fish or imprison the skipper.

I find it reassuring that simply protecting some parts of the sea for their own sake and the sake of the fish is being seen by more and more scientists who understand all the complexities of so-called fisheries management as the way ahead because it gets round incompetence, political egregiousness and intellectual dishonesty by their own kind.

CHAPTER 15
MCMEALS FOREVER

CLICK ON MCDONALD'S US website and you may learn more than you might wish to know about what goes into a 150-g (5-oz) Filet-O-Fish®. That is, unless you want to know about the fish itself. You are told that the battered fish in the bun is pollock or hoki. McDonalds does not explain that the pollock is walleye pollock from Alaska and that hoki is one of the cod family from New Zealand, or what the proportion of hoki is to pollock. Nor does it tell you how either were caught or whether the fishery is well managed. McDonald's 'nutritional facts for your McMeal' assume you want to know that the batter on the outside of the fish is made from bleached white flour, water, modified corn starch, yellow cornflour, salt, whey, dextrose, disodium pyrophosphate, sodium pyrophosphate, sodium tripolyphosphate and cellulose gum – presumably in case you have an allergic reaction to one of these ingredients and need to seek medical help. There is no explanation as to why anyone might want phosphates in their batter anyway. You are left with the impression that McDonald's website is written for litigious, nutrition-obsessed neurotics with a strange lack of curiosity about the provenance and whys and wherefores of what they are eating.

I can't help being curious, though, about what McDonald's means by a 'filet'. It is sometimes referred to as a 'fish filet patty', which raises suspicions that what we are dealing with here is mince (which it isn't). Minced fish is what makes up the lower grades of processed white fish in the fish fingers and breaded shapes sold in supermarkets. There is actually nothing wrong with it. Sometimes minced fish has a higher fat content and is therefore superior in taste. Americans, however, like their fish to taste bland (Europeans prefer a fishier taste). So fish for the American market will have what processors call the 'fat lines', the lateral lines of darker-coloured meat, taken out. This also improves freezer life because the fat lines can go rancid even when frozen. Fish fillets for the American market are called 'deep-skin fillets', having been skinned to leave just pearly-white flesh. For fun I tried asking McDonald's what was in their fish filet patties via the customer services email address – rather than through the press office, which is for privileged members of the media. No reply. The person on the other end of my email clearly thought there was no need to respond to an importunate foreigner.

Having talked to some of the people in New Zealand, Seattle and Alaska who supply McDonald's, I can reveal that what is actually in a Filet-O-Fish® is a deep-skin fillet, but not necessarily what the average person might think of as a fillet. McDonald's, which is known in the processing trade for having such demanding standards that some processors find it too taxing to deal with them, uses fillets frozen into blocks. These are frozen at sea or in an on-shore processing plant in Alaska. At the secondary processors, where they go to be battered, they are sawn up into fillet shapes while still frozen, so each Filet-O-Fish® may contain parts of several actual fillets. The fish 'fillet' thaws a bit during secondary processing when the warm batter is applied, but is immediately refrozen. It is then sent on to restaurants, where it is cooked. Compared with the fish sold in chip shops, which is iced but not frozen at sea and might have been out of the water for three weeks before you eat it, McDonald's fish is actually very fresh, having usually been out of the sea only a matter of minutes before it is frozen. The freezer life of a pollock block for most food manufacturers is a year after primary processing. (Freezer life depends on fat, which depends on species. The 100-year-old orange roughy, for example, is virtually indestructible in the freezer.) Funnily enough, the McDonald's website doesn't tell you any of this.

But none of this is leading where you might expect, and certainly not where the uncommunicative McDonald's would expect a book such as this to lead. For at least one of the fish used by McDonald's in the United States, the hoki, is certified by the Marine Stewardship Council (MSC – original slogan 'Fish Forever', now 'The Best Environmental Choice in Seafood'), an organisation that gives an independent certification of sustainability, an eco-label, to fisheries. The Alaskan pollock, which McDonald's uses in 90 per cent of the 275 million fish sandwiches it sells in the USA and Canada each year, is approaching the end of a two and a half year process of certification. Hoki is one of only eight fish stocks in the world that are MSC-certified eight years after the organisation was set up and began to investigate what made a sustainable fishery. The most surprising thing is that McDonald's are making so little marketing capital out of their use of the one whitefish stock in the world (hoki) that is actually certified as sustainably caught. This is a rare case of Ronald McDonald being knowingly undersold. This surprised me until I found out that McDonald's would have to pay royalties for the use of the MSC label.

The very idea that McDonald's, hated by greens and foodies, could make capital out of being on the side of the angels – and that their customers were therefore more virtuous than the denizens of exclusive restaurants – has caused shivers of revulsion among the righteous in the USA's environmental groups. They have just noticed that hoki has an MSC certification and that Alaskan pollock is in line to get one. Now a coalition of green groups is cutting up rough. They complain that the hoki should never have been given its certification. They accuse the Alaskan pollock fishery, the largest remaining whitefish fishery in the world, of being run on the basis of the discredited concept of maximum sustainable (or sustained) yield (see Chapter 7). They accuse the Alaskan pollock fishery, in the words of one of their campaigners, of 'strip-mining the ocean and treating the fish like a crop of corn in Iowa'.

The man who said that, whom I have come to know, was Ken Stump, a Seattle-based, conviction-driven environmentalist, who lives in a modest bungalow and writes heavyweight reports. Ken asked me: 'Can you really have your fish and eat them too, as the MSC suggests?' The inference was clearly that you could not. I thought this was exactly the right question to ask, but I found the arguments put by US environmentalists, like Ken, baffling from an ocean and a continent away. Had they unearthed something of substance that was actively wrong about the fisheries they were complaining about, or were they just whingeing for the sake of it to get the publicity that swells the coffers of not-for-profit organisations? I decided to find out, and booked a stop on my round-the-world ticket in Seattle, the home of the Bering Sea and Gulf of Alaska pollock fishery.

First, I'm afraid, we must go back a little to understand how big the stakes actually are in this battle between the fishing industry and a group of environmentalists on one side and another lot of environmentalists on the other. In 1995 one of the world's largest buyers of fish, Unilever, and the world's largest conservation organisation, then called the World Wide Fund for Nature (WWF), came independently to the conclusion that the world's wild fish stocks would not last long unless drastic action was taken. Mike Sutton, an American then based with WWF International in Godalming, Surrey, remembers the WWF concluding that it was never going to get the action it wanted by traditional advocacy, in other words, lobbying a lot of governments all at once, a process like herding cats. He discovered by chance that Unilever, the largest fish buyer in the West and owner of such brands as Birds Eye, Findus, Knorr and Iglo, was also becoming alarmed at the extent of overfishing, which they considered a serious risk to their frozen fish business. Mike arranged lunch with Simon Bryceson, a consultant to Unilever, at the Groucho Club in London to compare notes. There they discussed creating an alliance between business and environmentalists to certify fisheries, thereby giving the consumer the power to choose fish from well-managed stocks. The idea wasn't original – a Forest Stewardship Council already existed, intended to combat similar problems of greed and poor governance that were the death of primary forests. But persuading the fishing industry to trust itself to an alliance with environmentalists was a different kettle of fish.

There was nothing wrong, in theory, with using certification as a tool. Certification has existed for over a century and, broadly, it works. Teachers, doctors, nurses, policemen, auditors, engineers, ships' pilots and lab assistants require certification to practise their trade. Respected environmental certification goes back to 1946, when the Soil Association, the organic farming body in Britain, was born. Certification companies, like firms of auditors, exist all over the world and are used to buying in required skills. The only trouble Unilever and the WWF might not have foreseen was that the people who staff certification companies tend to communicate in impenetrably dull technical language that may or may not reassure the consumer.

At Unilever Simon Bryceson persuaded Catherine Whitfield, a native of maritime Canada who had experienced the cod crisis at close hand, that working with the WWF was in the corporation's interest. Anthony Burgmans, a tall Dutchman on Unilever's board, who later became co-chair of the company, liked the idea. He signed an agreement with the WWF at a ceremony in The Hague. Not long after that Mike Sutton phoned journalists to say that a new organisation was to be formed called the Marine Stewardship Council, which would organise the independent auditing of fisheries. The idea we journalists used to 'sell' the story to our editors was that this might reflect poorly on governments, who at the time still professed to be managing their fisheries well.

Unilever made some unusual promises for a major transnational company. It promised in 1996 that it would be sourcing all its fish from sustainable sources by 2005. Nearly ten years later it says it will fail to fulfil that promise because of the relatively small amount of certified sustainably produced fish on the market, but it still expects that three-quarters of its fish will be certified as sustainably managed by then. This, of course, is an admission that rather more than a quarter of this massive company's fish comes from sources that are presently driving down fish populations. It would be easier to be hard on them if all their competitors were not, by definition, worse. Unilever does seem to have done its best not to be caught out. Meanwhile, it has asked all its suppliers to confirm that their fish is legally caught and that they were not involved in species threatened with extinction. Unilever claims to have stopped doing business with those suppliers who could not offer this kind of confirmation. It certainly does not buy cod from the North Sea, and has sold its tuna-canning business. It has also asked the fisheries it deals with – Norwegian cod and saithe, Chilean and South African hake – to seek MSC certification. This, a pollock fisherman in Seattle told me admiringly, is what you call leadership. He could very well be right.

Eight years on the now independent Marine Stewardship Council – WWF and Unilever withdrew once the organisation was set up and running – allows only eight stocks to carry its label: western Australian rock lobster, Thames herring, Alaskan salmon, New Zealand hoki, Burry Inlet cockles, southwest UK hand-lined mackerel, Loch Torridon nephrops (also known as scampi or langoustines) and South Georgia toothfish. A long list of fisheries – topped by Alaskan pollock, the largest fishery for human consumption in the world, with catches of 2 million tons a year – is undergoing assessment.

It seems that the MSC has begun to have a real influence. When I interviewed Brendan May, its chief executive, he was cock-a-hoop because he had just heard that the producers of Norwegian cod, the major competitors to Alaskan pollock, had applied for certification. Chilean and South African hake were jostling to get in. 'The market is beginning to drive the process of conservation improvement,' he announced triumphantly. What appears to have led the scramble for certification among the world's whitefish producers is that those who had achieved certification had been seen to benefit financially. New Zealand hoki, he told me, had increased its sales in Europe 13-fold following certification. According to a recent poll, two-thirds of British consumers say that ethics influence their buying decisions, with the 18–35 age group most likely to base their purchases on green or ethical criteria. Research carried out for the MSC, Brendan May told me, had shown that the organisation could exert a powerful influence on world fisheries, making them more sustainable, even if it concentrated on getting market penetration for its label in Europe alone. This was because Europe sources fish from every ocean. The fishing industry and processors were looking to certification now, said Mr May, because 'certification is about allowing people access to markets which in ten years' time they won't be able to access, particularly in Europe'.

On the down side, the MSC has had more of a struggle getting its labels taken up and used in the United States and Asia. Even if all the fish awaiting certification do get certified – which is possible, as there is a pre-certification phase in which no-hopers, to spare embarrassment, are privately told they stand no chance – only 4 per cent of the world's fish supplies will be certified as sustainably managed. This is a similar proportion to the amount of forests currently certified by the Forest Stewardship Council, and while it could be influential, it is clearly not enough to save the world's fish. But Brendan May says it is entirely possible that about 40 per cent of the wild fish in European freezers and refrigerators could be carrying an MSC label by 2010, representing a real choice for consumers and exerting a powerful influence on the world's fisheries. Exciting stuff.

Just as the MSC – meant by its founders to be the only global eco-label for fish – stands on the verge of a breakthrough with the biggest players in the fishing industry, it has run into problems. One is competition from initiatives with different aims and labels, generally accepted by both industry and environmental groups as less rigorous. The other and more serious problem is criticism from environmental groups, and this brings us back to Seattle and Ken Stump of Alaska Oceans Network. Stump suspects that the London-based MSC is an exercise in greenwash by major multi-national fishing companies, led by Unilever. Even the WWF in the USA put in a 120-page complaint about the Alaskan pollock fishery, though they stopped short of calling for it not to be certified. There are plenty of reasons to suspect that the criticisms originated in petty jealousies of the 'not invented here' variety among US environmental groups, who are quite capable of their own forms of imperialism. The critics then raised points of substance, which deserve examination.

There are plenty of initiatives competing to harness consumer buying power, particularly in the United States. Some excellent seafood guides – one produced by the Monterey Bay Aquarium, another produced originally by the Audubon Society and now by Carl Safina's Blue Ocean Institute, both on the web – tell consumers what not to buy. These are the USA's equivalent of 'The Good Fish Guide', produced by Britain's Marine Conservation Society. There are several fish eco-labels in the United States, one of which, Eco-Fish, was in trouble on the front page of the *Seattle Post-Intelligencer* the day I arrived to check out the pollock fishery because a chain of wholefood stores had been using its label, erroneously, on tuna.

There are differences, though, between Europe and the United States. Where European environmental groups have focused on eco-labelling as a guide to personal choice, those in the USA have organised boycotts. SeaWeb, a non-profit organisation based in Washington DC, and Natural Resources Defence Council, organised a successful boycott of the Atlantic swordfish, which they called off when they perceived a stiffening of resolve to reduce quotas and tackle illegal fishing by ICCAT. The National Environmental Trust organised a campaign to persuade consumers to 'Take a Pass on Chilean Sea Bass'. Gerry Leape, a seasoned environmental campaigner and the organiser of that campaign, came to Europe as the leader of an alliance of groups, including Greenpeace, who challenged the proposed certification of Alaskan pollock and South Georgia toothfish by the MSC, and complained that the certification of hoki should not have happened. He said that bird and seal deaths caused by trawlers had been overlooked in granting the hoki certification, and the fishery was not in as good health as the certifiers said. His arguments about the Alaskan pollock were all about the endangered Steller sea lion. This was all rather difficult to check out as it was taking place on the opposite side of the world.

Leape, a genial fellow, told me he was generally in favour of the MSC and eco-labelling. He accepted that it was the only global label with buy-in from both industry and environmental groups. He simply wanted the MSC to take on his criticisms. There was a threat in the background that he, Greenpeace and others might reconsider their support for the MSC initiative if the MSC's board did not address their problems. This seemed fair enough coming from him. Coming from Greenpeace, who were late on the case in campaigning about the decline of world fisheries, it smacked of mischief.

Leape's initial submission had mistakes in it, which meant that some reporters in Britain did not take up his case. These he later admitted and corrected. He accused the MSC of having a board overwhelmingly controlled by fishing companies and seafood retailers (which wasn't true), conducting no annual audits (this was wrong) and being 'unable to say no' to a fishery. This was a tease, but it appeared true. The private pre-certification phase means that we have never been told who, if anyone, has been turned down (though Brendan May has said that North Sea cod and some tuna fisheries don't have a prayer). Nor have the MSC's certifiers yet turned down a single fishery seeking certification. That certification procedure is paid for by the fishery itself. These seem valid criticisms of the whole process. Leape went on to accuse the MSC of relying on national law as the standard fisheries should meet, instead of setting a higher benchmark based on consistent standards of its own. I had some sympathy with this complaint.

How, indeed, do you decide whether a fishery is sustainable? Can any fishery conducted by 90-metre (300-ft) trawlers, as is the Alaskan pollock fishery, ever be said to be sustainable? What is the acceptable level of attrition in other species or interference in the ecosystem? That was Ken Stump's question, if you remember, and a good one. To figure that out, we need to look at the three principles that certification companies retained by applicants have to consider when deciding whether to grant an MSC certification.

A fishery needs to score 80 per cent or more under each of the following principles.

Principle 1. A fishery must be conducted in a manner that does not lead to overfishing or depletion of the exploited populations, and for those populations that are depleted must be conducted in a way that demonstrably leads to their recovery.

Principle 2. Fishing operations should allow for the maintenance of the structure, productivity, function and diversity of the ecosystem (including the habitat and associated dependant ecologically related species) on which the fishery depends.

Principle 3. The fishery is subject to an effective management system that respects local, national and international laws and standards, and incorporates institutional and operational frameworks that require use of the resource to be responsible and sustainable.

The process cannot have seemed that difficult when the MSC certified the Thames/Blackwater herring, a smaller version of the North Sea herring, which spawns between the rivers Thames and Blackwater in the east of England and is caught in small drift-nets and landed not far from where I live, at West Mersea on the Essex coast. A total allowable catch of 128 tons is permitted each year. The MSC certifiers, meeting at the Company Shed, a rough and ready seafood café on the front in West Mersea, placed only one condition on the certification. The committee that regulates this small, inshore fishery had to find a way of coordinating all its vessels to ensure that fishing ended when the quota was used up. That was a piece of cake.

Little difficulty can have been involved either with the Loch Torridon langoustine creel fishery in the northwest of Scotland (there is a distinctly British flavour to the MSC's earlier certifications, as it was clearly cutting its teeth). The fishermen use fixed, baited traps, a vastly less damaging method than trawling. Creel fishermen found themselves in conflict with trawlers because – as the inhabitants of Arran have also discovered – Scotland removed a ban on fishing with mobile gear within 5 km (3 miles) of the shore in 1984. The Loch Torridon fishermen applied to the

Scottish Executive to have an area closed to mobile gear. They succeeded in 2001, when a closed area was established in Loch Torridon and the Sound of Rona, which is likely to bring many benefits, and not only for langoustines. The creel fishermen make a good income catching about 100 tons of langoustines a year, all of which are exported to Spain.

It was when the MSC certifiers got to one of the world's major commercial fisheries, for hoki, that things began to get complicated. The fishery is carried out by standard 70-metre (230-ft) freezer trawlers between the north and south islands of New Zealand and in the Pacific and the Tasman Sea. The fishery is conducted under individual transferable quotas, with draconian penalties for cheating, including the confiscation of quota, vessels and licences to fish. There was a question, though, about how much the stock had been fished in the past and how long it should be allowed to recover. There was also an annual by-catch of 1,000 fur seals and 1,110 seabirds, over 60 per cent of the latter being species of endangered albatross. An objection was lodged by a New Zealand environmental group, the Royal Forest and Bird Protection Society. It said that the certifiers had failed adequately to interpret and comply with the MSC's three principles. This had not happened before.

The MSC, then a new organisation that had not brought in the objection procedures it later established, set up an independent panel under Sir Martin Laing, a veteran businessman and patron of environmental groups, to consider whether a certificate should have been given. This said that the way the fishery was run was 'world best practice' – which may well be an indictment of the world's other fisheries – and confirmed its certification. But the panel was more sympathetic to Forest and Bird's view that little had been done to reduce the by-catch of seals and birds. The panel said that seal excluder devices should be tried. The MSC says its approach is 'one of continuous improvement, and it is not necessary that all problems be solved prior to certification'. Since the review panel in 2001, the certifiers' annual audits have criticised the Hoki Fishery Management Company for failing to complete sea trials of excluder devices – which appeared not to have worked – and has hit the fishery with some major 'corrective action requests', as the MSC calls its most urgent action points, which can lead to the decertification of the fishery. It does not look good for the hoki company either that it had to ask the New Zealand fisheries minister in 2003 to cut the quota because the hoki stock appeared to be in decline. Forest and Bird appear to have had a point about quotas being set too high, and may well be right that they still are. There is a suspicion, which the MSC should be trying harder to avoid, that the hoki fishery, once it got market access to Europe on favourable terms, has been making hay while the sun shines. Is the stock really being managed, in accordance with the MSC's Principle 1, for recovery? If the hoki stock declines further, the MSC will have some explaining to do, as will Unilever, which has placed its faith in the fishery as part of its path to sustainability.

While there are concerns about hoki, there seem to me fewer about the South Georgia toothfish. You can see why eco-labelled toothfish might have alarmed Gerry Leape, with his 'Take a Pass on Chilean Sea Bass' campaign behind him. But competition between unsustainable supplies and sustainable ones was exactly what the MSC was supposed to create, and it seems to me that a legal, sustainably managed fishery, provided its fish can be clearly identified by a 'chain of custody', should in time squeeze out illegal supplies in major markets at least. Eco-labelling is therefore less of a blunt instrument than a boycott. What matters is that the fishery is well run, well policed and that the stock is discrete, i.e. not part of a wider stock that is being poached.

An afternoon at Imperial College, London, with David Agnew, who designed the management system for the South Georgia government, convinced me that South Georgia toothfish are confined to the continental shelf around South Georgia. The other areas where toothfish are found – and poached like hell – are on the other side of a very large continent. The fishery is patrolled by the British vessel *Dorada*, with its heavy machine-gun, and surveyed by satellite and overflights from the Falklands. Around South Georgia the albatross and petrel by-catch is down to 20 a year because long-liners are required to set their lines under deterrent streamers and at night. Leape's objection, that Argentina disputes Britain's claim to South Georgia and the South Sandwich Islands, is not seen as an obstacle to the everyday business of international agreements by the world's diplomatic community.

Whether Alaskan pollock should receive certification raises perhaps the hardest issues, despite the current management of the fishery being, in the words of the assessors who spent two and a half years reviewing it, world class. The reason is the part the walleye pollock plays in one of the world's last functioning semi-wilderness ecosystems. Between 1970 and 1998 the pollock fishery in the North Pacific and the Bering Sea contributed 4–7 million tons a year to the world's supplies of fish. In the Bering Sea and the Gulf of Alaska catches are running at a staggering 2 million tons a year, all of which is caught by US vessels. The pollock is thought to have been significantly overfished on the Russian side, but how much this affects US waters is unclear. Since 1977 there seems to have been a major 'regime shift' in Alaskan waters. Pollock numbers have soared and the spawning stock is now as large as it has ever been. But Steller sea lions, an endangered species covered by the US Endangered Species Act, are 80 per cent down since the 1970s. Scientists and environmentalists blame the decline on fishing too close to the coasts where pollock, a main source of food for the sea lions, are found. Other scientists say that the case is not proved, and blame killer whales, which used to attack great whales and may have moved on to tasty young sea lions.

I visited the offices of At-Sea Processors Association in Seattle, who were funding the application for pollock certification. At-Sea Processors, who represent the large freezer trawlers, are strong on suits and lawyers. It is clearly a very profitable operation. I spoke to Craig Cross, who works for a company running 80-metre (270-ft) freezer trawlers – pollock fishing is a mid-water trawl with very little disturbance of the bottom – and he explained that the agreement in 1999 to allow fishermen to form cooperatives that divided up the quota worked in a similar way to a rights-based system. It had eliminated the race for fish, making the fishery much less wasteful and much more economically efficient. 'We'll spend a day looking for the right size of fish, where in the past you just kept fishing,' said Craig, who drove a smart Saab. All large vessels carry observers. John Gruver of United Catcher Boats, which represents smaller vessels of around 36 metres (120 ft) that land their catch in Alaska, said the fishermen's competitive instincts had been diverted from the race for fish on to by-catch avoidance. There is a large by-catch of salmon in the pollock fishery, which native Alaskans depend on for food, and the rules say you have to stop fishing when you reach your limit of salmon. 'The fisherman of the future isn't going to be measured by the fish he does catch, but by the fish he doesn't catch,' he said. I went away thinking that these were the most sophisticated and enlightened fishermen I had met anywhere.

But this is when I met Ken Stump of Alaska Oceans Network and heard the horror stories. There was a plausible link between sea lion decline and overfishing in the Shelikov Strait in the Gulf of Alaska, where the spawning population of pollock was around 2 million tons in the early 1980s and the sea lion population about 20,000. Now the pollock population was around 230,000 tons and there were only 3,221 sea lions. The two, of course, might or might not be related. There were no marine reserves in the Bering Sea or the Gulf of Alaska, which Ken was justifiably miffed about. The fishermen had also not mentioned the winter fishery for 500,000 tons of spawning pollock to provide roe for the lucrative Japanese market. These pollock, if left to spawn, would contribute to a much healthier population of fish. So this did sound like strip-mining the oceans. The only point on which I disagreed with Ken was his dismissive view that the MSC was a 'promotional gimmick of big industry players and had little to do with protecting the marine environment'. It seemed to me that this was too flip. In the USA the environmental agenda has made little headway in most major companies. I like to think Ken was underestimating how much in Europe, where the MSC was born, it has become part of corporate agendas. Anyway, what other solution to the problem of making fisheries more sustainable was he offering? Was another stand-off between environmental groups and two multi-nationals –

McDonald's and Unilever – going to lead to an improvement in the fishery or make it worse?

The MSC certifiers eventually scored the pollock fishery over 80 per cent against all three principles and recommended that it be awarded a certification, although subject to fairly stringent conditions. These, if applied and acted upon, would seem to drag the pollock fishery some decades into the future and to be a victory for Ken Stump and his kind on several counts: studies will have to be carried out to demonstrate that what the industry calls its 'harvesting' strategy is precautionary, even in the event of 'regime shifts'; the fishery will have to take ecosystem considerations into account when planning where fishing effort should take place; it will also have to improve assessments of the impact of fishing on the foraging habitat of the Steller sea lion. No certification will be given while the fisheries remains in breach of two domestic laws, one of them the Endangered Species Act. But (a black mark against the certifiers) there was no requirement to look into the eventual establishment of marine reserves or protected areas.

A criticism of the MSC is that it is impossible for a lay person coming cold to the assessors' two hefty reports on the organisation's website to make head or tail of whether they have done their job properly. At least a quarter of what they have written is opaque jargon that has no business in a public document, except in an annex. It is not written in any sense to give you confidence in the fish you eat, only to convince a paymaster, which happens to be a major industry association, that all the Is and Ts have been dotted and crossed. Large parts of it are fisheries science gibberish. Its competence is neither evident nor persuasive. It has to be taken on trust, which was precisely the approach of so much of the government science and policy that has left the oceans in their current sorry state. The MSC ought to know better. You can tell why people get suspicious of its intentions.

I myself was so annoyed with the MSC reports on the pollock fishery, having been through the ordeal of reading them, that I thought I would do a little journalistic digging to see if I could find out anything they didn't know, which would puncture the

professional pomposity of their certifiers. I did discover a little
detail that didn't seem too sustainable. Neither the MSC nor the
environmental bodies ranged against them, who claimed to know
far more than I did about the fishery, appeared to know about it
until I pointed it out to them. It is this. In Alaska the waste from
the vast amount of pollock that is processed to make all those
McDonald's fillets and Unilever fish fingers, together with waste
from other fish that get caught with them and the 'trash fish' that
nobody has a use for gets boiled down to make oil. This oil,
amounting to some 30,000 tons a year, could easily be used in
human fish oil supplements, in food for farmed fish, or to feed
animals, thereby reducing the amount of small fish (such as the
overfished blue whiting) that need to be caught for that purpose.
Instead, the oil is burnt – about 98 per cent of it, Scott Smiley of
the University of Alaska's fisheries technology centre in Kodiak tells
me.

The main reason for this absurdity is the Jones Act, a piece of
protectionist US legislation dating from 1924. This says that any
vessel sailing between two American ports must have an American
crew, something that makes transport between Seattle and Alaska
vastly expensive – and makes fuel expensive, for it has to be shipped
with American crews from Seattle, not the low-paid multinational
crews now commonplace on merchant ships. Scott Smiley tells me
that the fish oil, which is free to the plants and has a calorific value
only fractionally lower than kerosene, is used to generate electricity.
'It's a double win for them because of the transportation cost of fuel
from Seattle,' says Scott Smiley. Until the fuel is taxed, or the Jones
Act repealed, this obscenity is likely to continue.

I pointed this out to a staff member of the MSC in Seattle, who
had put me in touch with Scott Smiley, and she agreed that it was
'ludicrous'. In the MSC's defence, she said that what happened to
waste was not generally part of any certification. When you are
talking about a certification for sustainability, that's pure sophistry.
Don't I remember MSC Principle 3 requiring 'institutional and
operational frameworks that require use of the resource to be
responsible and sustainable'? Burning fish oil, measured against any

criteria you like, isn't responsible or sustainable.

So should the pollock fishery get an MSC certificate? It comes down to two questions. How good is the fishery? And – dear old Rashid Sumaila's killer question – what would improve it? The answer to the first is about seven out of ten, but with a little effort it could be an eight – good enough. What would improve it? What would mean, for example, that there was more chance of something being done about the Alaskan fish-oil scandal, certification or not? To me the answer is clear. Certify the fishery. If you think, as the US environmentalists appear to, that we should only certify perfection, what is going to spur improvement? Do environmentalists really believe that the commons are safe in the hands of a government that has so far failed to ensure the pollock fishery even complies with the Endangered Species Act? If not, what on earth are they whingeing about? I just don't understand why they are trying to ensure that only perfection is certified. It would be an act of despair not to certify the pollock fishery, warts and all. Then, of course, the warts will have to be removed.

So should my conscience be bothering me if I eat a McDonald's Filet-O-Fish®? If you are talking about my self-obsessed, litigious, nutrition-neurotic conscience, those phosphates with long names might trouble me and so might the calorie count. But if you are talking about my real conscience, about what is going on in the world and how I might influence it, I would advise the stick-thin patrons of exclusive restaurants selling endangered species to walk out and get their skinny asses round to McDonald's.

CHAPTER 16

BURNING THE MIDNIGHT OIL

Third Fisherman: Master, I marvel how the fishes live in the Sea.
First Fisherman: Why as men do a-land, the great ones eat up the little ones.

William Shakespeare, *Pericles* Act 2, Scene 1

ESBJERG, the largest port on Denmark's North Sea coast. A musty smell wafts into the town square on most days, particularly when the wind is coming from the direction of the giant 999 fish-meal plant down on the docks. Workers at the factory lower a large vacuum pipe into the trawlers' holds to suck up the silvery sand eels, sprats or pout. Billions of little fish are then carried into a massive industrial process that squashes, mashes and boils them, then divides them into oil and fish-meal. The meal is blended with other supplements to make food pellets for farmed pigs, chicken and salmon. The oil goes partly into the pellets for salmon, the rest to make human food supplements, margarine, paints and varnishes.

The first time I went to Esbjerg, over a decade ago, it was to investigate rumours that the trawlers supplying the plant were catching so many fish that there had been a glut of fish oil. The factory manager, an open and diligent fellow, had found a new market for his surplus of oil: he was selling it to power stations. The oil had nearly the same calorific value as fuel oil, with the added advantage that it was not taxed. It made poor-quality coal burn extremely well.

In Denmark, where they take a particularly utilitarian view of the sea, a minority was alarmed by the practice. In Britain and over most of Europe the burning of fish oil was regarded as an obscenity. Denmark's fisheries minister was forced to slap a tax on fish oil used as fuel to persuade the 999 factory to stop selling it for this purpose.

The sand eel, which is at the bottom of the North Sea food chain, is the fish that nearly everything else eats. It lives in the mud for half its life, and the factors driving its incredible fecundity have never been properly explained. My story about what the Danes were doing caused an outcry among salmon and sea trout anglers in Scotland, where catches were in decline. It upset bird organisations concerned about the number of unexplained mass bird deaths around the North Sea. Scottish trawlermen were stirred, too, because they have long accused the Danes of hoovering up juvenile herring, haddock and cod as by-catch in their fine-mesh nets. Those who catch white fish also knew that if the North Sea's cod and haddock stocks were to revive, the sand eel fishery would have to be reduced. Its huge abundance had resulted from the overfishing of fish that used to eat them.

Today is a dispiriting time for the 999 plant and the trawlers that supply it. But it is difficult to be too sympathetic. All catches are down, and workers are being laid off. In addition, the 999 plant's customers – Scottish, Faroese and Norwegian fish farmers – have been hit by a large drop in sales since researchers found high levels of cancer-causing dioxins and PCBs (chlorinated compounds) in the farmed salmon. The contaminants were concentrated in the salmon's food, which was made by boiling down small fish from Europe's polluted waters. The levels were such that the scientists recommended eating only one portion of farmed salmon a month.

The pollution dates from the days, only a decade or so ago, when the North Sea was treated by industry as an open sewer and dumping ground for toxic chemicals. Although now banned, these practices have cast a cloud over fish farming, which 'experts' still favour as the answer to feeding an expanding human population and to solving the problem of declining wild fish catches. By 2030, they say, aquaculture will dominate fish supplies. Who says that? Our old friend the UN Food and Agriculture Organisation. Given how wrong the FAO has been in the past – saying catches were going up when, in fact, they were going down – this statement is worth examining. When you do, you find it to be an observation of existing trends, not a reflection of what should happen or what people want – rather like Golden Delicious once being the most popular apple in Britain because it was the *only* apple you could get. The FAO is simply observing that fish farming is the fastest growing form of food production in the world – growing at 9 per cent a year and by 12–13 per cent in the USA. Nobody is asking us whether we want this: it is just happening. The continued destruction of mangroves in poor countries to provide shrimp for people living in rich countries is simply the market operating in a vacuum untroubled by ethics. It is a reflection of what will go on happening if we do not find ways of exercising any choice in the matter.

I am not opposed in principle, as many environmentalists appear to be, to fish farming as a process, even an industrial process dependant on factories such as the 999 plant in Esbjerg, if it can be run within sustainable limits. I happen to think that if I were hungry enough, I would be happy to eat farmed fish every day. If I were poor enough, I might well rip down a mangrove in Vietnam or central America to farm tiger prawns, assuming I knew the right cocktail of chemicals to keep them alive. But it is the responsibility of those of us who are fortunate enough to exercise choice to ask whether this is the way the world should be going.

I think that we, as consumers of fish, have to start with what we want, then modify this with the reality of what we are likely to need. What we want, all the trends show, is more fish, preferably tasty shellfish, such as big tiger prawns, or oily fish containing plenty of omega-3 fatty acids. What we need, if there are no wild fish or they are too expensive, is farmed fish. But what we definitely don't like, if you look at the public reaction that has followed the study in

Science, are fish that threaten to compromise our health by containing dioxins, PCBs, or residues from antibiotics and pesticides. We seem less troubled by fish farming that damages the environment – perhaps we see it as a necessary price to pay – but this may also be because we know very little about it. On the few occasions that people are asked, they say they would like fish farming to be sustainable, but this may be very difficult. Let us examine first the trends in fish farming, then the constraining factors, and try to see where we, as consumers, have any influence.

There are two kinds of fish farming, or 'aquaculture' as jargon-writers and people who want to neutralise the connotations of intensive farming like to call it. The first was developed by the Chinese at least 2,000 years ago and entails feeding waste vegetables to fish in a pond. The fish are herbivores or omnivores, and chosen for their tolerance to fairly murky water. When people talk about the growth of low-tech fish farming in developing countries for subsistence, they are talking about this. There are fewer problems with keeping vegetarian fish, such as tilapia. Domesticated strains grow faster than the wild variety. It is an option the developed world should consider doing more. The problem is one of taste. The taste for carp, for example, has pretty much died out in western Europe, but not in central European countries, such as Poland and Hungary. Medieval monks imported carp into Britain from the Continent and kept them in ponds so that they had fish to eat on Fridays. In Britain the declining taste in carp probably has to do with King Henry VIII deciding to kick out the monks.

What the West thinks of as fish farming, however, is the kind that has grown up in the past 30 years, which relies on feeding processed food made of ground-up wild fish to carnivorous fish, such as salmon, trout and prawns. The growth figures have been little short of staggering – shrimp (as Americans call prawns), mostly produced in the world's poor South, is now the most popular seafood in the United States. (I have never seen a food writer mention this, but I happen to know that to be imported into the USA all shrimp has first to be washed in chlorine bleach to kill bugs. What this does for the taste, I do not know, but I think we should be told.) Salmon is now the third most popular seafood in the United States. A decade ago Costco, one of its biggest retailers, did not stock fresh fish. Now it sells 15,000 tons of farmed salmon fillets a year. The cost of mistakes, mainly disease caused by keeping wild animals in close proximity with farmed ones, but also escapes and over-expansion, has meant that not everyone has made their fortune.

The whole carnivorous fish industry depends on the continued availability of small, wild fish, such as sand eels, to grind up into pellets. That is why there is trouble on the horizon, for the amount of small wild fish is finite, and so is the amount of fish oil and meal. Of the two, the determining factor is fish oil, which contains all the omega-3 fatty acids that we and salmon thrive on. This is because 70 per cent of what is currently produced is already used in fish food. On the other hand, only about 34 per cent of fish-meal currently goes into fish farms, 29 per cent into pigs, 27 per cent into poultry and 10 per cent into various other human and animal foods. Dr Stuart Barlow, director general of the International Fish-meal and Oil Organisation, tells me he warned the fish feed industry five years ago that fish oil was likely to be a limit to their growth. He warned that if they didn't find ways of substituting vegetable oil for fish oil, and to a lesser extent vegetable protein for fish protein, the world was going to run out of fish food by the end of this decade.

Remarkably, the amount of fish oil and meal produced was static from 1950 until the end of the 1990s. There are hiccups in supply every time there is a major El Niño warming event in the Pacific, when the reproduction of the Peruvian anchoveta tends to drop off. The Peruvian anchovy is the subject of the biggest industrial fishery in the world. In future, as the price rises, it is thought that the other uses of fish-meal and oil will drop away, and virtually all fish oil will go into fish feed. Dr Barlow points out that if people are concerned to cut waste, they should take their fish oil in capsules rather than it going back into fish food. Provided the supplier can assure you that the PCBs and dioxins have been taken out, that is an alternative to eating farmed salmon – a bit like cutting out the middle-man.

So is fish farming the solution to the decline of wild fish in the oceans and the search for more human food, or is it just a further problem? In theory, fish farming holds the solutions to some of the problems of wild fish. It is unquestionably the solution to the possible extinction of bluefin tuna in Europe caused by uncontrolled catches for the fattening cages. The Japanese have 'closed the circle' with bluefin tuna by breeding it in captivity, but this is much more expensive than catching it in the sea, which needs to be banned or heavily regulated by the Convention on Trade in Endangered Species. If the Japanese like bluefin tuna so much, let them pay the price. The Chileans have just closed the circle on the Patagonian toothfish, so poaching toothfish may be about to go the same way as poaching salmon, which was a major source of revenue in many parts of Britain only 30 years ago, when there were prolific quantities of wild fish.

The key question underlying the sustainability of any farming operation for carnivorous fish is where all the small wild fish that go into the feed are coming from. The static nature of the global industrial catch conceals the fact that we have been 'fishing down the food chain'. Forty years ago we were pulping the North Sea herring to make pig food, which contributed to the herring's collapse. Now there is a decline in sources of smaller fish, such as sand eel. This means few places are safe for some of the world's other stocks of unloved, unattractive but ecologically important fish.

We have already found out that there are grotesquely unsustainable catches of blue whiting currently coming from the North Atlantic on vessels such as the *Atlantic Dawn*. The vast majority of this goes into making oil and meal. Scientists are warning that with catches of nearly 2 million tons a year, this deep-water fish could be wiped out in two or three years, just like the North Sea herring. I checked with the records centre in Iceland – because they keep the best records – and I was told that 276,000 tons of the national catch of blue whiting was going to make fish-meal. Only 8,900 tons was kept for human consumption, mostly in eastern Europe, where consumers remembered how to cook blue whiting from a time when it was caught by the grossly subsidised Russian and Polish fishing fleets.

Catches of blue whiting for fish-meal raise questions that no one seems to want to discuss – that haven't been properly discussed since the 1950s and 1960s, when the British criticised the Danish and the Norwegians for catching perfectly palatable herring and turning it into pig food and fertiliser. If farmed salmon need three times their body weight in food made from other fish – cost cutting and waste reduction has taken that down to 1.1 times their body weight on the best-run salmon farms – then why don't we just eat the blue whiting? It has far fewer residues of PCBs and dioxins because it has not been boiled down and concentrated, and because it comes from the comparatively unpolluted waters of the mid-Atlantic. Blue whiting has absolutely no antibiotic residues. It also contains none of the pesticides that are used to kill the parasitic sea lice on the salmon, and is free of the flesh colourant – banned for use in terrestrial food – that is used to make the salmon's flesh pink. The only difference is that, like cod, it concentrates oil in its liver, not in its flesh. Blue whiting is not an oily fish, so if you want your omega-3s, you would be better off eating herring, which is in reasonable supply. Blue whiting is described by Icelandic chefs who have tried it as extremely palatable, rather like a small cod, for which they have some of the best recipes.

Horse mackerel, another of the world's staple industrial fish, *is* an oily fish. It is prized by the Japanese, and has graced at least one emperor's breakfast table. I have never eaten it, so I can't tell you what it's like, but even sand eels are said to taste nice fried up like whitebait. So what is stopping us eating blue whiting, horse mackerel, sand eels or even Peruvian anchovy? The market. That's us, the consumers. So instead of asking for farmed salmon the next time you're at the fish counter, try asking for blue whiting or horse mackerel and see what happens. You might just start something. The market is fickle and can be used to do good.

I am told that the creation of a demand for table-quality blue whiting would mean that the trawlers would not be able to fish at the speed they now do with their vast Gloria trawls. Catching low-value industrial fish means packing fish into the cod end of the net, where it becomes mangled and crushed. This is perfectly good for 'reduction', as pulping it for fish-meal is known to the experts. Fishing for human consumption would mean catching fewer fish of higher quality and higher value. Only the salmon farmers would be unhappy.

The blue whiting gold rush is not the only aspect of industrial fishing that raises ethical concerns. I am indebted to the sharp eyes of Monica Verbeek of the conservation group Seas at Risk for pointing out a disquieting entry in the Danish industrial catch recorded by the International Council for the Exploration of the Sea. It appears that 4,195 tons of round-nosed grenadier caught in the Skaggerak Strait found their way into Danish fish-meal plants in 2002. The round-nosed grenadier is a fish that first reproduces at the age of 8–10, lives to 75 years, and is many times more vulnerable to overfishing than the short-lived, highly fecund species that are normally thought of as 'industrial' fish. It is probable that this large tonnage of grenadiers was accidentally caught while fishing for something else, probably cod or ling. There are places where the species appear together and are caught by the deeper trawls – though, of course, fishermen trawl this deep only because there are no fish anywhere else. The round-nosed grenadier are therefore a by-catch that other trawlers might simply discard, but the Danes happen to have a convenient reduction plant. So, just maybe, this catch was merely an efficiently used waste that prevented some other fish having to be caught. The alternative is too horrible to contemplate. Directed fishing for deep-sea species for fish-meal would represent a descent into a new circle of hell, anarchy and stupidity for the world's fisheries. I am told that the Irish fleet has considered it.

If fishing for deep-sea fish to boil down is a descent into hell, then there is another circle of hell devoted to a new refinement of industrial fishing. This is what *The Economist* calls 'the diversion of low-value fish from the mouths of people in developing countries into the mouths of well-fed fish in the developed world'. I have heard of limited examples of this happening already: whole sardines and pilchards from the great Mauritania fishery off West Africa being fed to bluefin tuna in so-called tuna farms in Australia, which fatten tuna for the Japanese market. If this were to become widespread, it would amount to an obscenity on an imperial Roman scale.

If the pressures on the world's small wild fish are like this now, then what will they be like in a few years' time? It is no longer salmon, but halibut and cod that are now causing a wave of interest in northern and western Europe. Cod are being seen as the answer, in Norway and the EU, to cheap competition from Chile. Research has been poured into the technical problem of how to feed a fry that, unlike salmon, has no large yolk-sac it can live off in the early days of its life. The research seems to be paying off. In 2004 the huge fish-farming company Nutreco will bring 350 tons of cod to market. By 2020 the company expects to be farming 400,000 tons of cod in Norway, mainly for export to the now largely cod-free EU. The figure of 400,000 tons is almost the same as the quota for wild fish caught in the Barents Sea by the entire Norwegian and Russian fleets. I have not yet seen any sign that anyone is thinking of the possible implications of this massive expansion for wild cod. Would fish farms take the pressure off wild stocks, allowing them to recover to their original abundance, or is it more likely that farming would introduce genetic pollution and disease into the wild population?

What is proposed for cod in Europe sounds like a massive expansion of fish farming – and it is. It is nothing, however, compared with what is being contemplated in the United States, where, according to the Institute for Agriculture and Trade Policy, a Minneapolis-based coalition of consumer and environmental groups, fish farmers are working with the federal regulatory agencies to privatise parts of the open ocean for fish farms. Currently there are experimental off-shore fish farms situated 5–320 km (3–200 miles) from the coast – in Alabama, Mississippi, Florida, Hawaii, New Hampshire, Puerto Rico and Texas – growing high-value species, such as red drum, amberjack, summer flounder, cod, halibut, red snapper and cobia. Assuming that the technology exists to construct them without allowing huge escapes of domesticated fish, there doesn't seem yet to have been any consideration of what open-ocean farms might do to the sea bed in terms of pollution, or to wild fish in terms of pollution or disease. The Minneapolis-based institute appeared certain that the Bush administration would submit an offshore aquaculture bill to Congress in 2004 that would set up a policy framework for offshore aquaculture. Environmental groups in the USA tend still to argue for the 'public stewardship' of the sea, which has caused problems everywhere. They would be better off acquiring some rights, or helping fishermen to acquire them, and using them to stop fish farms getting a foothold.

There is a conflict between wild fisheries and aquaculture that has a long way to run. In the United States there is talk of using genetic modification to insert a gene into salmon to make it grow bigger and faster. In many other countries there is a belief that genetically farmed fish are a step too far. The concern is about what happens if the fish escape and breed with wild fish, as they are almost certain to do unless they are kept in contained facilities on land – and even then the security of these cannot be guaranteed. Disastrous introductions of alien species have occurred around the world, from the poisonous cane toad in Australia, introduced to kill pests of sugar cane, which now kills a whole range of native wildlife from snake to crocodiles, not to mention domestic pets, each year to the Atlantic jellyfish in the Black Sea, which ate up all the zooplankton and replaced the anchovy which previously formed a valuable commercial fishery. Both should stand as examples of why GM fish are not a good idea.

This is not meant to be a book about recreational fishing, so I will touch only briefly on the cautionary tale of salmon farms on the west coast of Scotland – an example it is worth thinking about carefully before undertaking the kinds of expansion of fish farming that are being contemplated. All my life I have been a sea trout angler, and the west of Scotland should be my Mecca. The first time I went there to fish, as a teenager, I went to Loch Maree, then the best sea trout loch in Scotland. In the mid-1970s, at least 2,500 sea trout migrated up the River Ewe and were caught by visiting anglers in Loch Maree. When I went there 15 years later, the Loch Maree Hotel, one of the most famous fishing hotels in Scotland, was empty of anglers. It had become just another stop on the tourist route. There were a few devotees who went out to fish for sea trout, but the annual catch was under 100, and one year numbered just 19. The same thing has happened to river systems all over the west coast and the Scottish islands because the estuaries and the coast have become infested with huge concentrations of sea lice as a result of salmon farms. Once the presence of a sea louse on the side of a fresh salmon was a cause for celebration – the sign of a fish fresh from the sea, as sea lice drop off in fresh water. Now it is a cause of foreboding.

For nearly a decade I had been writing stories about the mounting evidence that sea lice were responsible for the sea trout collapse that happened in the 1980s, just as salmon farming took a firm hold on the Scottish west coast. Most of the evidence came from Ireland, where scientists, such as Ken Whelan of the Salmon Research Agency, were quicker to report and more open about the problem. In Scotland, where salmon farming went virtually unregulated during its formative years, nobody in the establishment wanted to know anything that might detract from what was seen as a wonderful way of creating employment in rural areas. In this case, it was a form of employment that put others out of business.

Then, all of a sudden, the scientific position on sea lice changed. Professor David Mackay, an area director of the Scottish Environmental Protection Agency, told a gathering of his peers in Norway in 1999: 'The case that damage to stocks of sea trout and wild salmon by sea lice associated with caged farming is very serious in certain circumstances has been made to the point that it should now be accepted as beyond reasonable doubt.' Three years later, the strongest evidence to date between sea lice infestation of wild migratory trout and fish farms was published by the Scottish Executive's own fisheries laboratory at Faskelly. It showed that the infestation of sea trout in Loch Torridon and the River Sheildaig was of farmed origin, and this was shown to be at its worst, i.e.

potentially fatal to wild fish, in the farmed salmon's second year in the pens. The WWF said it handed the salmon farmers a smoking gun. Allan Wilson, the Scottish minister for environment and rural development (a difficult combination of roles to pull off successfully), responded that it was 'crucial to strike a balance between the needs of this growing sector and its impact on the environment'. Striking a balance in this case meant doing nothing.

Two years later not a single salmon farm has been moved from the path of migratory fish. Scotland's home-grown anti-aquaculture campaigners continue to blow holes in the industry's credibility, and there is increasing pressure on the industry from all sides. The science is firm enough now, one would have thought, for a court case for damages to be brought by the owner of a fishery against a salmon farm, but none has yet materialised. A welcome development is that some fish farming companies, such as Marine Harvest, have begun to use their undoubted skills in farming fish to restock rivers, such as the Lochy, with sea trout. I shall believe that the fish farmers have earned their place as full members of society only when they restore Loch Maree.

All the same, fish farmers are not going to go away. Like other industries, they place their faith in technological improvement in what is still a very young industry. They make the point that many of the food scares, which Europe seems addicted to, are just that. The industry knew about dioxins and PCBs in European waters several years ago, and knew also that the European Commission was likely to introduce limits for toxic chemicals in fish food. Two years before the researchers' findings appeared in *Science*, based on salmon tested three years previously, factories such as the 999 plant in Esbjerg began treating the fish-meal to remove high levels of dioxins and PCBs. Had researchers tested Scottish salmon in 2004, rather than three years ago, the results would have been very different.

Graeme Dear, managing director of Marine Harvest in Scotland, assures me that the fish oil shortage, too, will be solved by innovation. Nutreco, Marine Harvest's parent company, now believes that it can substitute 75 per cent of the fish oils in fish feed without any ill effects for the salmon, or presumably for the person eating it. Ulf Wijkström of the FAO talks about the 'fish feed trap' being solved by using 'as yet unexploited aquatic proteins', such as Antarctic krill: that will test CCAMLR's much boasted ecosystem management (see page 138). Ultimately, fish farming may not be that good for us, but it is better than having no fish at all – or fish that is unaffordably expensive. We will have to live with the fish farming industry, and the industry will increasingly have to live with us by paying more of its environmental costs.

Some salmon farmers have tried to deal with society's fears about intensive fish farming by offering organically grown salmon. This is a respectable try, but undermined by the fact that fish have to be fed with concentrated fish-meal that is made up of fish often caught in unsustainable ways and containing, as we are now told, concentrated contaminants. An attempt to devise a trustworthy labelling system for the least environmentally damaging farmed fish – such as the organic movement provides in agriculture – is earnestly to be wished for so that the consumer of farmed prawns, salmon and sea bass has a choice, and competition drives up standards. So far, assurance schemes are run only by the aquaculture industry itself. Meanwhile, the only organic fish in most people's minds are wild fish.

The scary thing is that nobody seems to be considering the impact on those wild fish of fish farming on the scale that is now being proposed on the coast of Norway or in the open ocean off the United States. Fish farming, even with conventional techniques, changes fish within a few generations from an animal like a wild buffalo or a wildebeest to the equivalent of a domestic cow. Domesticated salmon, after several generations, are fat, listless things that are good at putting on weight, not swimming up fast-moving rivers. When they get out and breed with wild fish, they can damage the wild fish's prospects of surviving to reproduce. Many of the salmon in Norwegian rivers, which used to have fine runs of unusually large fish, are now of farmed origin. Domesticated salmon are also prone to potentially lethal diseases, such as infectious salmon anaemia, which has meant many thousands have had to be quarantined or killed. They are also prone to the parasite *Gyrodactylus salaries*, which has meant that whole river systems in Norway have had to be poisoned with the insecticide rotenone and restocked. What would happen if there were large cod farms near to the largest concentration of wild cod left on Earth, in the Barents Sea? If Norway's idle regulators have their way, we could be about to find out.

Increasingly, we will be faced with a choice: whether to keep the oceans for wild fish or farmed fish. Farming domesticated species in close proximity with wild fish will mean that domesticated fish always win. Nobody in the world of policy appears to be asking what is best for society: wild fish or farmed fish. Were this question to be asked, and answered honestly, we might find that our interests lay in prioritising wild fish and making their ecosystems more productive by leaving them alone enough of the time. We might find that society's interests did not lie, as the FAO has led us to believe, in a massive expansion of aquaculture.

On land, incentives have already swung from intensive forms of agriculture towards extensive ones. Pigs in England are no longer raised in concrete silos, but outdoors with room to snout around and wallow in mud. Following the problems caused by BSE, terrestrial animals are no longer fed their own kind mashed up and rendered down. It is the fish farmers' nightmare that a BSE equivalent lurks for farmed fish. There are compelling reasons for favouring extensively grown fish rather than intensively grown ones. But where do you see that in any government policy? Everywhere in the sea the incentives and subsidies favour the killing of too many wild fish and the building of more fish farms.

This is the opposite of what society says it wants. Just as nobody asked whether it was a good idea to take a million tons of sand eels from the North Sea and boil them up in the 999 factory, it just slowly happened. What will happen in future is what the aquaculture industry thinks will make money, unless we make an active choice to arrange things differently. We do have a choice as to whether the incentives driving the aquaculture industry – lax regulation and often subsidies – should exist. We do have the choice of restructuring wild fisheries so that they are no longer a way of buying the votes of fishing communities, but of managing fish stocks properly so that they provide health and nutritional benefits to society. We would be healthier and eat fewer chemicals and pesticides used in fish farms if we were to have a clear hierarchy of priorities, with wild fish at the top and farmed fish at the bottom.

We need to remember that the conservation of wild fish is a human health issue.

CHAPTER 17

THE THEFT OF THE SEA

NOBEL HOUSE, Smith Square, London, 1997. I was talking to a senior official in charge of fisheries in an oak-panelled room at the then Ministry of Agriculture, Fisheries and Food (since transformed into the Department for the Environment, Food and Rural Affairs). It was spring, and outside the window I could see budding plane trees in the Embankment gardens that run down to the Houses of Parliament – a safe, sovereign English scene. I asked him what it felt like to play off the demands of the fishermen each year against the needs of the North Sea cod, which scientists were already warning was in danger of collapse on a Newfoundland scale. He admitted that it was like backing towards the edge of a cliff in the fog. You knew the edge was there somewhere. You could hear the sound of waves crashing and the cries of gulls wheeling below, but you did not know when you would topple over. I finally understood then what I had often suspected: that Europe's fisheries were out of control.

If the sound of the waves and the cry of the gulls were loud then, they are much louder now.

Back in 1997 the man sitting in front of me was revealing, consciously or not, that he had no more than a marginal influence over events, despite being in command of his brief and having a mighty bureaucracy behind him. The wheel-house of this great 'trawler' bent on destruction of Europe's fisheries was in Brussels.

Seven years later, following the lightning tour of some of the world's fisheries that I undertook to write this book, I knew I would have to go to Brussels to run the story to earth. While it has tried to highlight the problems of global fisheries and global markets, this book was written in Europe, so all it can ultimately hope to offer is a European's view of the situation. The world has shrunk, and fish caught in foreign seas have become everyone's business, but in the end, wherever we are, our understanding of other people's political systems and problems will always limp behind those of our own. What I *can* do, however, is show how Europe became one of the biggest and most malevolent spiders in the world's marine food web.

So let's assume that we are sitting as jurors in the court of world opinion. I will now read the indictment against the European Union for its handling of its common seas and its negligent control of fleets which fish other oceans. Ahem.

Paragraph 1: Two thirds of Europe's main commercial fish stocks, including cod, hake, plaice and sole, are outside their biological safety limits. This means they are on the verge of collapse. Some spawning stocks, such as the west of Scotland cod and cod in the Kattegat channel between Denmark and Sweden, may already have collapsed – just like Canada's northern cod – only Europe hasn't faced this yet.

Paragraph 2: Anarchy prevails in Europe's common sea because fishermen have no faith in the system, or that the system will catch them. Spanish fishermen catch 60 per cent of their hake illegally. Scottish fishermen, frustrated by declining incomes, which they put down to Europe's poor management of fisheries and not their own greed, threatened recently not to obey European quota rules. What they appeared to be proposing would make little difference, since half of what they land is illegal anyway. Instead of fishing illegally as usual, they would be fishing illegally as a protest. Europe's enforcers, who have never appeared to do much, now appear to have given up. The penalties are not enough to deter: they do not reflect the economic gains to be made by fiddling.

Paragraph 3.1: Europe is incapable of copying perfectly good rules that other countries have and making them work. Two examples: in New Zealand, if a vessel's skipper is convicted of under-reporting catches or misreporting his position when he caught them (the latter checked against the satellite record), his quota – a tradeable possession of some value – is confiscated forever, his licence to fish withdrawn and the vessel impounded for good by the taxpayer.

Paragraph 3.2: Another example of Europe failing to follow the example of best practice from around the world is shown in its failure to ban the discarding of dead fish. Cornish fishermen are throwing perfectly good monkfish over the side because they have too little quota and can't, as in other countries, buy more, or land the fish and give the money to charity or fisheries research. The European system dictates that undersized fish, capable of being big fish next year or the year after, must be returned to the sea, dead, where they do no good to anyone but crabs: there is no requirement to stop fishing and move on, and no observer programme to make anyone do so.

Paragraph 4: Europe's fisheries policy is destroying other forms of life in the common sea, and damaging natural resources important to other people, with impunity. Dolphins are being slaughtered off the Cornish coast, and harbour porpoises are meeting untimely ends all over the European seas. Cold-water corals have been protected, but since satellite vessel monitor data doesn't seem to get used by the enforcers, one wonders if this is not a grand empty gesture. Stocks of bass, important to bring in recreational anglers, who are a significant part of tourism on the Cornish Riviera, are being scooped up by industrial trawlers for little financial gain. This happens because the law of the sea is biased in favour of commercial fishermen, and nobody is really responsible for the day-to-day running of the fishery.

Paragraph 5: Having ruined its own fisheries, Europe is now exerting a malign influence across the globe. Its growing demand for fish, comparable to Japan's, has created an appetite for canned tuna that now exceeds the USA's. European fishermen are wiping out threatened big-eye tuna in the Indian Ocean, the Atlantic and possibly the Pacific, and killing off the entire cast list of *Finding Nemo* at the same time. That's not all. Europe's fish farms are leading to the unsustainable pursuit of small fish, such as blue whiting.

Paragraph 6: Contrary to the spirit of the Doha Development Round of the World Trade Organisation in 2001, which says that fisheries subsidies should be phased out everywhere, Europe continues to subsidise fisheries to a disgraceful degree. Subsidised tuna-fattening operations in southern Europe have caused the overfishing of one of the ocean's most prized assets, the bluefin tuna, and are inexorably pushing it towards extinction. And Europe continues to spend hundreds of millions of euros buying its fishermen access to other people's waters, from the west coast of Africa to the Solomon Islands, in a gesture of naked neo-colonialism. These 'partnership agreements' do little for the poor countries in question and end up exporting Europe's illegal fishing practices and contempt for sustainability to other parts of the world.

Paragraph 7: European vessels stand accused of pirate fishing in most oceans of the world, yet there is at present no way of prosecuting them in their countries of origin for what they get up to elsewhere. In support of this allegation I call as witness a New Zealand sage of the fishing world, Sir Tipane O'Regan, former chairman of Sealord, the New Zealand fishing group which catches hoki in New Zealand and Latin America. He told me: 'Most of the major pirate vessels are owned by European capital – mostly Spanish or Russian – and on the whole the industry knows who they are.' The EU is not responsible for the malign influence of mafia-run government in Russia, but it is responsible for policing its own citizens.

In response to this list of allegations, surely the European Union can plead only guilty as charged? If there were a prize for the most disgraceful country or group of countries on Earth for pillaging the sea, the European Union would be the most favoured recipient. (Now there's an idea.) Romano Prodi, president of the European Commission, would have to stand at the ceremony and make an acceptance speech. As Brendan May of the Marine Stewardship Council put it to me: 'The European Common Fisheries Policy is a laughing stock in many parts of the world. People in New Zealand and Australia don't understand how Europe could get something that should be simple so badly wrong. The extinction of its own – and the world's – fisheries is being presided over by people who ought to know better.'

How could things go this wrong? How can the people who ought to know better let this happen? The answer, as always, is politics. What fisheries biologist Ransom Myers said of Newfoundland – that the system is set up to fail – is true in Europe too. A clue to the failure is in the title of Europe's Common Fisheries Policy (CFP). If we have seen anything in this book, it has been that seas managed as commons that no one is responsible for are the most likely to fail. Seas controlled by people who know who owns what and have legal responsibilities and assets are the most likely to succeed. Not all the commons fail, but giving rights and responsibilities to individuals or small groups of named people generally improves the fishery.

The problem with Europe's common sea is that it is managed as a common, and the only way round that problem is to enclose it so that everyone knows exactly who owns what, has a clear idea of what their responsibilities are and can take an interest in increasing the value of their assets, whether these are marine reserves or quotas.

There are many people in Britain, however, who think that the problem with Europe's Common Fisheries Policy is that it is European. These people include most of the fishermen's organisations and the Conservative party, which has decided that it would take Britain out of the CFP, draw a line down the middle of the North Sea and 320 km (200 miles) out beyond Cornwall and the Shetlands, and have the Royal Navy fire across the bows of any Spanish trawler that crossed into British waters. I have some sympathy with the 'Four Inch Shell Option', as I shall call it. But I must point out that it has little to do with the conservation of fish.

People who think it has should look at two things: the overfishing triggered in Newfoundland and Iceland when the 320-km (200-mile) limits were drawn, and what happened in the North Sea before the Common Fisheries Policy was formed by a regulation passed by the original six members of what was then the European Economic Community in 1970, to the horror of Britain which presided over some two thirds of continental waters and was about to join. British fishermen want the repatriation of Britain's fisheries because they want be able to catch all the fish, instead of the Spanish doing so. And they probably would. Fish would still swim across invisible lines in the ocean, and the conservation of their stocks, which are in a parlous state, would depend on agreement with other coastal states. Ironically, something rather like the Common Fisheries Policy would have to be invented. The only virtue of the 'Four Inch Shell Option' from the point of view of conserving fish is that if the people who believe in it got elected, it might create the institutional revolution that seems to be necessary for the whole system to be set up properly to conserve fish *and* fishermen.

What we actually need to do is to understand the reasons why Europe's Common Fisheries Policy, even the new 'reformed' CFP, is currently set up to fail. We are back to politics: Irish, Scottish, Danish, Dutch, French, Italian and Spanish politics, the last with Generalissimo Franco's long shadow not so very far in the background. The way that Europe runs its fisheries amounts to theft. The sea and all the livings things in it have been stolen from the people and given to fishermen, rather as the countryside was given to farmers in the years when Europe was trying to feed itself after the Second World War. That, too, caused ecological disaster.

Fishermen are represented by fisheries ministers, and it is ministers, not fishermen, who decide what is going to happen to fish. This makes the fishermen unhappy because they get a load of unworkable rules and no say in the proceedings, but the sea is nevertheless run in their interest by people who believe that, politically anyway, fishermen were sacrosanct. (This is rather like a previous generation of politicians called industry ministers, who believed that miners mattered.) Nobody else's interests get a look-in – not the citizen, of whom Europe has some 454 million (as of May 1 2004), not really the treasury's, not the marine environment's, not the anglers', not divers', not children paddling on the beach or watching dolphins from a ferry – just commercial fishermen, many of whom are now engaged in what is in effect criminal behaviour with two logbooks: one for legal catches, the other for illegal ones.

Environment ministers may cluck all they like about sustainability – as they did at the only joint environment/fisheries ministerial meeting about the North Sea there has ever been, in Bergen in 1997 – but the fisheries ministers still haven't got round to agreeing long-term recovery plans for declining species, such as cod, hake, plaice and sole, as they were mandated to do seven years ago. When dealing with matters of the sea, bureaucrats and politicians assume that the power is held by fishermen. In fact, it does not: it lies with the citizen, and if citizens in sufficient numbers don't like what is happening, the power could be taken back again.

The first principle of any true reform should be to make prime ministers, treasury ministers, environment ministers, representatives of people who aren't fisherman, take part in all deliberations about sustainability. At the moment fisheries are not really any of their business. Meanwhile, the only political influence over the sea's bounty is exerted by an industry – the wild capture fishing industry – which in Britain is roughly the size of the lawnmower industry. No one would dream of allowing the lawnmower industry to dictate the policy of a sovereign state or even a federation. The lunatics have taken over the asylum.

The fact that the sea is presided over by lunatics who believe there should be commercial fishing in 100 per cent of the sea breeds a culture that is corrosive. Two erroneous beliefs have been allowed to flourish. First, that you can cheat biology. Second, that you can keep people happy in far-flung communities in the west of Ireland, Scotland and Spain by allowing them to fish, when in fact the gallop of technology means that this year maybe only half a dozen people in the village can fish sustainably, and next year it will be four. Right now 20 of the rogues are at it.

Sooner or later, people will realise that their birthright, a healthy ecosystem, has been stolen and will want it back. By then the ocean is going to be in even worse shape than it is today.

*   *   *

Brussels, January. The lovely Belgian capital of boulevards, bourgeois townhouses and civic buildings, which survived the ravages of world war relatively undamaged, is slowly being devoured by a fungus of monumental buildings needed to accommodate the civil servants and politicians of the federation, or as some still prefer to think, the trading community, of which it is the capital. In May 2004 the European Union expanded by ten nations to 25, with the accession of several East European states, led by Poland, together with the Mediterranean islands of Malta and Cyprus. I discover as I walk out for lunch with a friend who works at the European Parliament and his wife that the lovely 19th-century Gare Bruxelles-Luxembourg has been felled to make room for all the new bureaucracy – enlargement will, of course, mean more bureaucracy.

Things are different in Brussels – both more and less civilised than the Anglo-Saxon world I come from. In a city of so many languages, it is important to take time for civilities, and lunch is important as the social fulcrum of the business day. The European Commission, the permanent government of Europe, has a two-hour lunch break, but here it is an opportunity to eat well and choosily, rather than to eat a lot. The question in my mind, however, is whether the chances of a wrong being righted have improved or not as a result of new countries joining the EU. As they include parts of the former communist world and reunite the Austro-Hungarian empire (a byword in its day for sclerosis – as the EU's fisheries and agriculture policies are today), I am not optimistic. Brussels is turning into a mirror image of the political culture on the other side of the Atlantic, once described as a demostasis – a culture in which every interest had a counter-interest, so nothing ever changes.

The EU, to be sure, is not particularly good at changing things that have gone wrong. Or if it is, it does it at a glacial pace. Maybe, I muse, this is because most of Europe is now formed of cultures that were set up in a big bang after defeat, reconquest or revolution. What, I wonder, is going to be the big bang that can save Europe's oceans and the global fisheries over which it now holds sway? Attempts by civil servants from cultures more used to muddled, incremental change, such as Britain to change the Common Fisheries Policy do not appear to be working very quickly. The European market and the European electorate include some of the most informed, green and progressive people in the world, and Europe's potential for doing good by its purchasing decisions is considerable – as the MSC has shown. But its political structure, and its deference to fishermen over sustainability, is unreconstructed and holds it back.

It is 3.00pm, the time European officials schedule their first appointments after lunch. I scuttle off to the faceless offices of the directorate general responsible for fish (DG Fish, as it is known) in Rue Joseph II. There I put it to Harry Koster, the charming Dutch head of inspection, that a superstate like the EU has tremendous potential for bringing about a rational use of the sea (just what Michael Graham, the sage of Lowestoft wanted), and not merely in its own seas, but in oceans across the world. Harry's smile stops me. He corrects my terminology with a voice steeped in Brussels political realities. 'We are not a superstate. We are a federation. We have to cooperate together. With living marine resources you have to cooperate together. Only coastal states can do the job. What you see is the coastal states making a big mess of things.' He says that what was needed came in the 2002 reforms – a European inspectorate to crack down on the coastal states with their devolved regions that are making a mess of things, to give hot pursuit across jurisdictions and to even out the disparities in enforcement around the EU. I noticed that a roomful of consultants, including British academics I know, left his office as I went in and return to his office as I leave.

I reckon that Harry is trying to set up his new baby, the beefed-up European inspectorate, so that it does not go native in its new headquarters in Vigo, the world capital of illegal fishing.

The message you get from European officials is that things are getting better, albeit very slowly. As Harry puts it: 'People have to be sensible. If you look at the future, you get totally frustrated. If you look to the past, you are surprised by what you have achieved.' You have to believe that if you work in DG Fish, or you would go mad – as mad as the system. But it is the job of journalists like me to hold up the status quo for examination: Harry told me frankly that every other cod and 60 per cent of hake were illegally caught. He thought an acceptable level of illegal fishing was perhaps 2–3 per cent. It's therefore an even chance that the cod on your plate is stolen – from you and me. It is also the job of scribblers to offer up proposed improvements from Europe's politicians and regulators to the public so that they can compare them with a more ideal world, in this case, with the ocean in its former prolific state 50 years ago, or what it might actually be like if a proper settlement were made, with common agreement, to restore that abundance. This I shall attempt to do.

I find the same guarded optimism in the office of Eskild Kierkegaard, a Danish scientist who is now principal administrator for the conservation of stocks. It is the same optimism expressed by Franz Fischler, the fisheries commissioner from the land-locked country of Austria, who, to his credit, has brought the language of sustainability into fisheries and tried to make the member states, particularly those in the south, face up to the fact that they are presiding over a spiral of decline.

Every year a committee of the ICES delivers its fish stock assessments – in a very obscure document, published on the web, which should in fact be properly written in the community's languages, then bound and published for the public to read, as assessments are in Iceland. For the years 2002 and 2003 the ICES said that fishing for North Sea cod should be banned and that fishing for any fish that led to a by-catch of cod should be drastically reduced. You will recall that David Griffith, the director general of the ICES, issued a statement expressing frustration that the amount of spawning cod in the North Sea in 2002 was down to a mere 38,000 tons, the tonnage of a single car ferry.

Yet for the past two years, just before the council of ministers met for their annual horse-trading session to decide the quotas for the year, Franz Fischler has come out with his own proposals, which ignore the ICES's recommendation for a ban. Why? If Mr Fischler believed the cod was in a poor state and conservation measures were not being properly enforced, as he had been quoted as saying, why didn't he say that cod fishing should be banned and threaten to resign if ministers didn't do so? That was what the men in Canada's Department of Fisheries and Oceans (DFO) should have done when they found out how low cod stocks were in Newfoundland, according to Lesley Harris, chairman of the independent inquiry into why the northern cod collapsed. Instead, he just sniped at ICES in October 2003, saying their recommendation of a zero catch for cod almost everywhere and for whiting in the Irish Sea was 'rather stark'. One suspects that Mr Fischler clearly enjoyed his position, his comfortable office and his even more comfortable EU salary and expenses too much to take a stand of this kind.

Eskild Kierkegaard's reply to the question 'Why not take the scientific advice?' is as honest as it is simple: 'It would have meant you had to stop fishing. Politically that was impossible to achieve. The council of ministers would never have adopted a moratorium in community waters. It would have led to the collapse of the processing sector. We can't have the fleet in port for ages. The amount to pay for them to stop fishing would be very big.' Kierkegaard says there is still far too much fishing, and the main problem is the control of landings. But he also says, possibly because he believes it, but more likely because he is an EC official, that the package of measures just agreed by ministers, including the new inspectorate, should begin to deal with it. Ministers had just signed up to a management plan for cod and other species, which aims at a 30 per cent increase in the stock each year. 'We believe we see the light at the end of the tunnel,' he added. I didn't have the heart to crack the old joke that it might be the light of an oncoming train.

There is nothing to say that the 'aim' of restoring stocks will work, just as the 'aim' of not crashing them has not worked for the past 15 years under European management, as stocks have sunk lower and lower. Ignoring the scientific advice is no way to deal with fisheries problems, as we have seen from around the world.

The best practice in these desperate situations include paying fishermen not to fish, buying many of them out once and for all, or creating mechanisms whereby the market can buy them out by allocating rights. By not resigning, or threatening to resign, Mr Fishler made sure none of these things happened.

John Farnell, the most senior British-born official in DG Fish, gave me a disarmingly wide-ranging interview that lasted well into the evening. His spin was the same: we've turned the corner. He said that the 'living dangerously philosophy was beginning to be understood as being dangerous' by ministers. Decommissioning schemes had led to sizeable cutbacks in the Scottish whitefish fleet, and made them a more realistic size. People further south in Europe, though, were not facing up to reality.

He had a pretty frank appreciation of those realities:

This is an area where the technology has moved very fast and is permanently moving. The industry must get smaller every year. Because of technology, shrinkage is necessary to keep this sector in balance. The Norwegian minister recently told some villages in the north, 'If you want to survive, you must accept there will be fewer people in the village fishing every year.' If a Norwegian minister says that, *we* should have the courage to say it.

How did he justify priorities over the whole common sea, theoretically the property of Europe's citizens, still being set by an industry that, in some countries, was no larger than the lawnmower industry? He smiled and hesitated. 'This is an industry which is locally very important in areas that don't have a lot of alternatives – in Scotland, on the Irish coast and in Spain. It is difficult for governments to take the decision of closing that down in a short time.'

Mr Farnell then reviewed the positive side of the balance sheet. After a decade of resistance from fishermen, and because of the real fear that stocks were collapsing, the EU had finally (and after a decade of being urged to do so by the commission) implemented a system of effort controls in the North Sea and North Atlantic. This was miles better than its disastrous total allowable catch and quota system that depends on fishermen not cheating. It had set up a system of regional councils that included fishermen, which might, eventually, lead to the management of the North Sea, for instance, by those who lived around it. This would exclude outsiders, such as the Spanish, who have different agendas. The EU had abolished subsidies – unless they were for safety equipment. (He was conveniently forgetting subsidies to fish in other people's oceans and the subsidies to tuna farms, but that is another department.) And Franz Fischler has finally been putting pressure on the Spanish, French and Italians to declare a 320-km (200-mile) limit in the Mediterranean, that other European sea, where fisheries management has long been noticeable by its absence. (Spain and France are in favour, but Italian fishermen, who plunder the shores of Croatia daily, are the big losers. Greece also has 'problems', presumably because it does the same to Turkey.) Doing something about the rampant overfishing of tuna, swordfish and so on was possible only within the context of a 320-km (200-mile) limit.

He almost had me convinced that things were going to get better. Then I asked him the commons question – 'Did the Common Fisheries Policy have to be common?' – and I was amazed by the answer. 'We don't exclude the idea of moving towards individual fishing quotas or property rights. I'm an advocate of market forces and property-based assets myself. As far as this has already happened in the Netherlands, we believe it increases everyone's interest in the rules being complied with.' Well, indeed it does, but wasn't it possibly the key to slightly more than that, to the fishermen having a financial interest in leaving fish in the sea? Farnell replied carefully: 'The Community view of how to do things stops at the point of dividing up the cake. We are in the game of getting best practice known about and picked up by member states.' Mr Farnell did not say so, but I think he knew it: the establishment of a system based on rights and individual responsibilities, and not on commons, would amount to a fundamental political reform of the way we deal with Europe's common sea.

Blinking slightly at John Farnell's generosity with his time – I entered his office just after 6pm and did not leave until just after 8pm – I emerged scratching my head and trying to answer the old question: was the European Commission the problem, or the solution?

The answer, of course, is both, I reflected on the plane back from Charleroi to Stansted, a flight that cost £19.99 one way compared with £99 on the less environmentally damaging Eurostar train because of the clout the aviation industry exercises in continuing to pollute the commons. The commission has the expertise and the people passionately committed to solving the problem of fisheries, but they are frustrated by a political accommodation that means they can never succeed in reforming a system that was set up to fail.

It is the flabby accommodation between political institutions that is wrong in Europe. The ministers should propose, the commission should dispose. It is, in fact, the other way round. The commission makes perfectly good suggestions for a system that would actually work, and they are watered down into something that won't work by fisheries ministers. This would not be a problem if the commission were set up like a national government (what is scary is that it *is* the government of Europe), with cabinet ministers who all had a say in policy and an effective veto, unless matters were brought to a vote. The problem with government in Europe is that all negotiations with industries and interested groups, such as environmentalists, are handled effectively within sovereign baronies or fiefdoms controlled by each commissioner. So the environment commissioner, who causes no end of trouble for Europe's manufacturing industry by restricting the amounts they are allowed to pollute right down to the smallest nanogram, has no influence, effectively, over any sustainability question in the oceans. Finance and trade commissioners are too grand and unenlightened to be interested. This is a disgrace. The environment commissioner can insist that something is done if fishing affects a non-commercial species, such as dolphins or cold-water corals, but even then the commission and other councils of ministers do not necessarily have to do much about it. The European Parliament seems to consist entirely of MEPs representing various interest groups whinging about the handouts given and rules made by the commission and ministers, but observing the same absolute divisions between the different baronies' responsibilities. This is the accommodation that has grown up. It is in no EU citizen's interest, and must be unpicked if the citizens are to reclaim the sea.

Do we need to go to all that trouble if what Mr Farnell and his colleagues say sounds so sweet and reasonable? Yes, we do. I worked out why when I got home and compared my notes of what John Farnell had said – in his explanation of the commission's line that things have turned the corner – with what Harry Koster had told me about how it would all work. Farnell told me everything would be all right now because Europe's fishing ministers had accepted effort reduction, the first rung on the ladder to controlling what fishermen do in the sea. But when I went line by line with Harry through how it would work at sea, it was clear that the new system wouldn't work much better than the old one. Harry Koster told me that the commission wanted to fix the number of days each fishing vessel was allowed to fish and how many days it had to spend in port. 'Everyone [i.e. all the member states] rejected days in port because it was transparent.' The ministers decided instead to fix the number of days fishermen were allowed at sea by gear, i.e. according to the kind of net used: 120 mm (4¾ inches) for cod, 80 mm (9¼ inches) for prawns, etc. You might think somebody would have consulted Harry, the person whose job it would be to enforce things in this way. Oh, no.

Harry said: 'How do you control gear? You need an army to control these things.'

The commission wanted ministers to monitor the days at sea by satellite vessel monitors, but all member states were against. Harry's verdict: 'It is a federation where members [who put on a public face of doing the rational thing that will control fishing] can hide behind each other. The ones who do not want it [effort reduction] are the fishermen. They are dead against. They prefer total allowable catch and quota because they can fiddle more easily than they can do with effort.'

The whole problem suddenly stood out in all its simplicity. The rules were made *so that they could be broken.* The European system was, and is, set up to fail.

We have seen already that the only thing that will restore the seas to the state of abundance they enjoyed at the end of the Second World War is another six-year rest. Since the stocks are now in a far worse state than they were in 1939, even six years may not be enough. Anything else is just bumping along the bottom. Without a plan that stretches beyond the next three years, and has as its (perfectly possible) aim the restoration of fisheries to their former abundance, how is one to respect any reassurances given by European officials or politicians?

So the system must go. The only question is, will it be the British way, incrementally by painful but rational decision, possibly as a result of the sensible suggestions of Mr Blair's Strategy Unit, or by some kind of fall-of-the-Berlin-wall-style revolution?

Before we decide which of these is more likely, I will tell you, very briefly, what I would put in its place – assuming that Europe's 'barony problem' were sorted out.

My plan to save the sea

In the better-governed parts of the world there is a person known as a fisheries manager. In Australia, for example, he or she actually manages the fishery according to the law, with a board of maybe four people, including a fisherman or an environmentalist. It is the manager's responsibility that the rules are observed, that things are run well and that there are plenty of fish in the sea. He is like the land agent on a well-run estate. If there is cheating, he goes to the government and asks for it to spend more money on the police, or gets the estate owner to do so. If he comes across anything that is against the law or his principles, he will resign and go to run someone else's fishery. If things go wrong, he is out. If he is doing his job properly, the quota at the end of the year will be worth more because there are more fish in the sea and he will earn a bonus.

In Europe's common sea, there is no one person who is responsible for cod. All changes to quota that become necessary and all management measures take at least a year to agree because they have to go through the ministers' annual horse-trading session in Brussels – where ministers *actually fiddle about with the law every time for their own political ends*. This is lunacy. First, the aim of managing the common sea must be established properly – the restoration of the EU's fisheries to their post-war abundance. A body of law, like a constitution, then needs to be agreed for fisheries and, as far as this is ever possible in a democracy, set in stone. This law should include strict limits on the amount of spawning stock that can ever be given as quota *once that restoration is complete* (25 per cent and no more for cod, as in Iceland), the location of marine reserves, closed areas for fisheries management reasons, technical measures involving nets, and so on. Then the whole of the management process needs to be de-politicised. It needs to be taken off-line – removed from the political area and given to a competent manager to manage for a period of years, not unlike the euro. You would appoint a manager for the fishery by species, such as cod, or if this is too difficult in a mixed fishery, you would appoint a manager by area, such as the North Sea, the Channel or the west of Scotland. This manager should be appointed every five years by the whole commission – just like the director of Europe's central bank that runs the euro. His remit would be strict, and if he didn't fulfil it, his contract would not be renewed.

But first you would have to cut the fleet. There are two ways of doing this: the part-state, part-private way, or the very expensive state-only way. The least expensive way is to set up a system of individual transferrable quotas for fishing off-shore, listening to all the advice from around the world to make sure that the crucial first allocation is fair to existing fishermen, vessel owners and processors. Then it is up to those who get the quota to decide for themselves whether to stay in for the long term or to sell their appreciating quota as a pension, or to start another business. (I was delighted to find that a system of ITQs for the British fleet was one of the recommendations of Tony Blair's Strategy Unit report, which came out while this book was in production, but I see little merit in the recommendation that these quotas should last a term of only 15 years. What incentive do they think this will create for fishermen in year 14?) Some state money would be available on a strictly one-off basis to buy up the difference between current quotas owned by fishing vessels and the quotas thought sustainable – around 50 per cent of what are allocated now. A different system, tied to the owners of specified small vessels, would be given for fishing within 32 km (20 miles) of the shore. Fishermen convicted of any infringement or cheating would automatically lose their quota and licence to fish, and their vessel would be seized and scrapped (as in New Zealand). This would take care of technological creep, for the fleet would contract every time there was a conviction. In parallel with these reductions in the fleet, the commission would set up, on behalf of the citizens of Europe, a system of large marine reserves, extending to a quarter of the North Sea and an equivalent area in the north Atlantic within the 320-km (200-mile) limit. Fishing vessels would be banned from these reserves, and the right of passage would be allowed only in life-threatening emergencies.

In the pièce de résistance of my plan, Europe would allow the citizens themselves to police these reserves and the fisheries of the world where European vessels fish by making the satellite monitoring data on all EU vessels available for all to see on the Internet in real time. At present, such information is strictly guarded to protect the fishermen's confidentiality. The idea is that each fishermen's knowledge of the fishing grounds is his own intellectual property, his advantage over his competitors. Remove the race for fish from the system and make fish abundant and there would be no need for confidentiality. Open access to satellite data would be one of the conditions that went with the generous gift of fishing quota to fishermen by a benevolent public – who own the sea.

Full public access to satellite data from European vessels would be the pre-condition of any distant-water agreements signed by Europe on behalf of fishermen. This means we could all turn on the computer and see, for example, just what the *Atlantic Dawn* was up to. So-called 'partnership agreements' would be negotiated by the EU, but paid for by the fishing industry. Private arrangements between EU companies and other states would be illegal. Companies based in the EU would be made liable in the EU for fishery crimes committed on any ocean of the world.

That, in summary, is how the citizens of Europe could reclaim their common sea and help to police the oceans of the world. The chances of all this happening, of course, are zero – unless one of the following things happens.

Scenario 1: People and their representatives begin to realise what is going on beneath the waves and want to organise things differently before any irrevocable barriers – such as the extinction of cod – are crossed. Tides in public opinion do change. Europe finally lost patience with its farmers for laying waste to the countryside and converted production subsidies, which caused environmental damage, into more neutral subsidies with some benefits for the countryside and wildlife.

Scenario 2: There is a cod crash, like Newfoundland's, and enough clamour is made about it. The nightmare suspicion I have is that the cod have already been written off in the corridors of Brussels because closing the North Sea, a mixed fishery, is politically impracticable. When the end comes the fishermen will simply shrug and move on to the haddock, without anyone asking the people of Europe what they wanted – to put an end to fishing for six years, or to abandon all hope of catching cod forever?

Scenario 3: The people who believe in the 'Four Inch Shell Option' get elected in Britain. In this case, there will be a short political window in which things could get interesting.

Until then, I'm afraid, Europe will go on backing towards a cliff in a fog.

CHAPTER 18

RECLAIMING THE SEA

I pointed to the Surrey bank, where I noticed some light plank stages running down the foreshore, with windlasses at the landward end of them, and said, 'What are they doing with those things here? If we were on the Tay, I should have said that they were for drawing the salmon-nets; but here –'

'Well,' said he, smiling, 'of course that is what they are for. Where there are salmon, there are likely to be salmon nets, Tay or Thames; but of course they are not always in use; we don't want salmon every day of the season.'

I was going to say, 'But is this the Thames?' but held my peace in my wonder, and turned my bewildered eyes eastward to look at the bridge again, and thence to the shores of the London river; and surely there was enough to astonish me . . . The soap works with their smoke-vomiting chimneys were gone; the engineer's works gone; the lead-works gone; and no sound of riveting and hammering came down the west wind from Thorneycroft's.

William Morris, *News from Nowhere, An Utopia*, 1890

LOWESTOFT, SEPTEMBER 2090. The herring is in. The sailing drifters, the only vessels allowed to fish on the edge of the town's marine reserve have been filling their boxes. The scene is reminiscent of a 17th-century Dutch painting – only with modern clothes, navigation technology positioning and safety equipment. The inshore-caught herring, a great bounty that has opened up again since the reseeding of the Dogger Bank with herring spawn, is greatly prized, and the restored seafront market is full of tourists who want to buy some from the inshore fishermen who catch the 'silver darlings' in traditional hemp nets.

Some tourists go for the whole experience on offer and go out at night with the herring fishermen who make a good living from the business that is now called heritage fishing. Others go to restaurants to have the herring fried in oatmeal by the chef and to eat oysters from areas carefully re-seeded out at sea. Wild oysters were reseeded in the great experiment of restoring the Dogger Bank around 2040, and have been building up steadily, but the law says they may only be fished for, like the inshore herring, from sailing vessels. Locally smoked kippers are in season and fetching good prices in the fish market, but the best prices of all go to the huge hand-lined soles caught on the edge of the 'no-take' reserve. The East Anglian newspaper says that there have been so many herring and anchovy off the east coast that the bluefin tuna, which have not been recorded off the English Coast since the 1950s, have been spotted in the reserve off Scarborough.

A few hundred metres along the coast children in wetsuits have been pursuing a giant skate all morning, watched over by the great-grandson of one of the last Lowestoft trawlermen – a warden in the town's marine reserve. The skate has no fear of people, and consents to be stroked before billowing away across the bottom. It can head away for miles in each direction quite safely, for bottom gear is banned in this section of the North Sea. The glass-bottomed boat has been following it, too. You can see quite far enough, though the visibility is only about 10 metres (33 ft). The water has been clearing over the past five years thanks to the build-up of oysters on the bottom. The other great excitement on the edge of the reserve today was a pod of bottlenose dolphins, playing among the wind farms out on the edge of the reserve and then playing tag with the returning drifters.

Captain Nemo's restaurant on the restored seafront is doing a roaring trade in fish from the inshore boats, these include species like red mullet which would not have been thought of as North Sea fish a few years ago. English white wine, now grown as north as

Yorkshire, goes well with the fresh, salty fare. The younger clientele arrive in wetsuits after sailing or swimming. The menu is full of local fish, sole, catfish, haddock and whiting caught on regulation jigs and hand-lines. An alternative supply comes from the haddock and witches (deep-water sole) caught by the 20 or so quota-trawlers in the northern North Sea fishing grounds. The trawler fishermen accuse the inshore fishery of being a theme park, but it is actually providing something that people want and rich nations can afford – a way of being part of nature and the maritime heritage. It brings in tourists, creates jobs and has given new life to the town. All the new building on the seafront is causing the city elders a headache in keeping up with the planning conditions, particularly since sea levels are a metre higher now and the amount of land protected by sea walls has shrunk. But the town has turned around in two ways since the early part of the century, economically and physically. All it once had was a much-reduced port and the Norfolk Broads, to the back of the town, which have lost their lustre since the creation of the Great Sole Reserve.

What children growing up today think of as entirely normal, we as educated adults know to be a great achievement. In the 20th century and the early years of this century the world's oceans were in a sorry state. Europe in the 20th century gradually dealt with the problems of pollution that had plagued the continent in the 19th century, killing off the salmon in rivers and making canals and waterways so poisonous that you had to have your stomach pumped out when you fell in them. But the problems of capture technology in the oceans had been underestimated and have taken nearly another century to tackle. Even then, the reforms worked only in the advanced democracies and in the regional organisations that are based in them. There continue to be problems around nations with enormous populations, such as Indonesia, the Philippines and, of course, along the coast of Africa, where the great up-welling fisheries of the west coast are now as much of a memory as its rainforests.

In Europe, though, we finally cracked it. It took the near-extinction of the bluefin tuna from the Mediterranean, and the total extinction of the cod from the North Sea, to achieve. Forty years later we are still attempting to reseed the North Sea with cod from Norwegian farms. It is possible that the climate, or the ecosystem, may have changed too much for them ever to return to their former abundance. But the variety of fish in the sea has if anything increased. The present-day arrangements of inshore fishing licences being limited to the least damaging methods and the offshore trawlers being run by a few companies with a financial interest in leaving fish in the sea dates from the 2020s, as does the network of large marine reserves that began to restore the oceans abundance, and were seized on by politicians as a popular measure and therefore grew.

It was almost certainly the Oceans Summit in 2012, 20 years after the Earth Summit in 1992, that set in place the laws needed to govern the middle of the oceans, but the reform of inshore waters in Europe and the creation of reserves actually has it origins in the period 2005–12, when a few reserves began to be created by law. There was a great dawning of public awareness about the oceans at the time of the Oceans Summit, which led to the oceans becoming the responsibility of the deputy prime minister in most European governments. Until then, the fishing industry had held sway over the sea.

Paradoxically, the few people who do fish are much wealthier than they were then, and so are the shore-based processing industries and the tourism industries which have grown up around the new, strictly limited forms of fishing. The sourcing of fish to the ocean, vessel and method by which it was caught became popular in the first decade of this century, and is taken for granted today. It was around that time that the premium for wild fish began and toppled some of the intensively farmed fish, such as salmon. Gradually many people developed a preference for some of the plentiful but overlooked fish which used just to be boiled up for meal. There's a restaurant on the front at Lowestoft, just past the statue of Michael Graham, that has specialised for years in recipes for sand eel and blue whiting.

*   *   *

Lowestoft, 2004. We can all dream. But when it comes to the oceans, we don't do it enough. The consequences of going on as we are hardly bear thinking about. In another 50 or 100 years all the world's oceans will be like the Mediterranean or the Java Sea – largely empty of everything but tiny fish attempting to evade the nets long enough to reach breeding age and mostly failing. The Mediterranean will be even more like a giant built-up swimming pool because the swordfish and bluefin tuna will have disappeared, as surely as the bison did from the great plains of North America. The last ones are bred using technological methods that have been forced on the industry by necessity. The great up-wellings of Africa will be almost gone in terms of their fisheries, with the exception of the southern hake off Namibia, which will fetch huge prices in Spain.

Some people in countries with growing populations near the coast will undoubtedly starve. The amount of fish off West Africa will be so small that Senegal and other countries once rich in fish will need food aid, just as Ethiopia and the Sahel do now. Fish farming will rule and poison even the open sea. The newspapers will be full of stories about diseases and mutations that farmed fish cause to the remnant wild fish populations. There will be more farmed fish than there is now, fed on soya beans and laced with genetically modified additives that mimic fish oil, but these will taste quite different from wild fish, which will fetch enormous prices.

To reverse these trends, we need to do several apparently simple things:

- Fish less. If we fished at about half the pace or less that we do now, the bounty of the oceans would grow and we could eventually harvest more.
- Eat less fish, or eat fish less wastefully caught.
- Know more about what we are eating, and reject fish caught unsustainably.
- Favour the most selective, least wasteful fishing methods.
- Give fishermen tradeable rights to fish, accompanied by new responsibilities.

- Create reserves that will cover the migration hotspots for the big game fishes, such as tuna and swordfish, on the high seas and 50 per cent of the entire area of intensively fished and used places, such as the North Sea.
- Make regional fisheries bodies responsible for the high seas work properly instead of merely monitoring the decline of the populations they are meant to preserve.
- In Europe, and within other 320-km (200-mile) limits, we must organise a quiet democratic revolution, whereby the citizens gain overall control of the sea.

Wanting these things enough will bring them about. Already the market in Europe is switching towards fish managed in the right ways. As yet there are not enough of them. But the pressure is already on the fisheries which are the least sustainable. They are most likely to be rejected by the inquiring reader of a menu, or the concerned consumer who knows a thing or two at the wet fish counter or at the freezer. The products of those fisheries, such as North Sea cod, are likely to find their markets dwindling among the major retailers – generally far more sensitive to customer preference and environmental campaigning than politicians. The fisheries of the future are going to have to come up with reasons why consumers should think they are sustainable.

Remember what John Gruver said, 'The fisherman of the future isn't going to be measured by the fish he does catch, but by the fish he doesn't catch.' To begin measuring what fishermen do we need to know what fish we are being offered to eat – what kind of tuna it is as well as how and where it was caught – and if we don't get told we should decline loudly, so others can overhear. An awful lot of work will have to be done in the fisheries of the world to produce these fish that we want to buy. The problem at the moment is that the examples of best practice in the ocean lie dotted around the world in little puddles, like the fast dwindling cod in the North Sea.

We have on offer two futures. One requires difficult, active choices starting now. If we don't take those choices, the other future will happen anyway.

CHOOSING FISH – A GUIDE

This book is not a consumer guide. At least that was not my primary intention. This has been more of a personal journey through the world's oceans, looking at fishing methods and their effect on target species and the rest of the marine ecosystem. On the way, though, I have come to realize just how much consumer choice can make a difference in changing the way that the oceans are managed and the fishermen pursue their catch. Therefore, I think, it would be a cop-out not to include the briefest and most personal conclusions about what to avoid eating, what to be cautious about and what to eat without a twinge of conscience.

The list is necessarily idiosyncratic. It defers to, but sometimes disagrees with, consumer guides produced by the Marine Conservation Society, the Monterey Bay Aquarium and the Blue Ocean Institute. These are, rightly, the first port of call for anyone trying to decide the ethics of what to eat because they are updated regularly and have input from a wide range of scientific opinion. I would recommend eating anything certified or in the process of being certified by the MSC (i.e. that has passed through the pre-certification process). I am less convinced by other certification schemes, none of which are as global or involve so many shades of opinion in the certification process. I recommend the IUCN Red List and Fish Base, both on the web, as a means of resolving after-dinner arguments about which species is which, what is fast or slow growing, vulnerable or less vulnerable to over-exploitation.

On the subject of information, I do not see why Iceland, a country with a population of little more than 300,000, can produce an accessible Blue Book each year on the status of every commercially exploited marine species from (controversially) whales to capelin and prawns, and the members of the EU and other major nations cannot. The IUCN's website, though much better than it was, contains the same information (I am told, though I can't always find it myself) but in far less accessible form, let alone one that encourages the citizens of Europe to hold their politicians to account.

Ultimately, consumers should be given enough information about fish to make their own decisions about what to eat. So lack of information is as good a reason for not buying something as any other. This means, for example, that I do not consider it acceptable to sell something called 'tuna' without saying which species it happens to be and how it was caught. As a matter of personal choice, I will now refuse to eat any fish that is not identified as to its species, ocean of origin and method of capture – which also means turning down things called 'Tuna crunch sandwich filler' and 'Italian tuna steaks' (was the tuna Italian or the way of preparing them?) By rejecting fish that is not properly labelled or sourced, consumers vote with their feet for a world better informed about what is going on in the sea.

So here, with profuse acknowledgements to all the above sources of information particularly the Marine Conservation Society, is what I would personally rather not eat – unless the management of stocks improves radically – and what I would choose if it were on the menu.

The dirty dozen – fish to avoid

Atlantic cod – from spawning populations recognized as overfished such as Newfoundland or the North Sea. Crime should not pay. At the moment, half the cod caught in the North Sea are stolen, whether from the sea or the people of Europe depending how you look at it. Icelandic cod, Faroese cod and maybe Norwegian cod are

OK, with the caveats made earlier in the book. Cod caught on rod and line or inshore by highly selective methods would be a reasonable alternative, too.

Atlantic haddock

Also listed as vulnerable by IUCN, though this pre-dates a certain recovery in the stock. Caught with cod, so North Sea haddock should be avoided and the best that can be hoped for is that fishing pressure on cod, haddock and whiting is cut drastically to give these species a rest.

Bluefin tuna

In the process of being disgracefully overfished in the Mediterranean, already disgracefully overfished and in need of a respite in the west Atlantic. As one of the oceans' astonishing and endangered mega-fauna, it is long overdue for a CITES listing. Environmental groups take note.

Caviar

Given the price of the stuff, most of us can feel virtuous avoiding wild-caught caviar.

European hake

Grotesquely overfished. Sixty per cent stolen in European waters. South African and Chilean hake appear to be more sustainable alternatives, and we wait for the MSC to check this out.

European sea bass

Appears to be heavily overfished, whatever IUCN ICES thinks. Trawl fisheries target spawning and pre-spawning fish and kill lots of dolphins. Fish would be far more valuable to the economy managed as a sporting resource. Rod and line caught or farmed fish are preferable, with obvious caveats about sustainability of farmed fish food.

North Atlantic halibut

Listed as endangered by IUCN. Exception to be made for line-caught fish from Icelandic and Faroese waters.

Patagonian toothfish (Chilean seabass)
Avoid, except MSC certified fish from around South Georgia and, where traceable, fish from now better-policed Australian waters around Antarctica. Inquire where it comes from.

Grouper
Many species are overfished. Consult FishBase.

Snapper
Many species are overfished. Consult FishBase.

Orange roughy
As with round-nosed grenadier, almost certainly unsustainably caught, with the possible exception of roughy from New Zealand. Damage to sea mounts a factor to have bad dreams about wherever they are caught.

Scallops
Pick the farmed variety or dived ones at a pinch. Absolutely avoid the dredged variety as dredging wreaks havoc on the sea bed.

Fish to think about
Plaice and sole
Overfished in European waters. Try over other flatfish such as flounder, witch (deep water sole) or dab. Some eastern English Channel sole is within safe limits and may get an MSC certification.

Swordfish
For a brief period, it may be OK to eat North Atlantic swordfish – check the assessments from ICCAT – but almost certainly not swordfish from anywhere else, certainly not the Mediterranean.

Sharks
Sharks of all species are long-lived and vulnerable to overfishing. Mostly caught as by-catch in catching swordfish and tuna, so there is an argument for directing concern at those fisheries but avoiding shark caught deliberately. Inquire how it was caught.

Tuna

Canned (purse-seined) tuna is scary for two reasons – the fact that it may contain endangered bigeye and involves an awesome, but unquantified by-catch. We urgently need to know more about how much by-catch there is and whether it can be avoided, perhaps by banning FADs. It is for Heinz, Princes and other big tuna canners to tell us. Until then, there is reason to feel deeply uncomfortable about eating a tuna sandwich. There is no reason to feel any happier that your can is marked dolphin friendly.

Long-lining for tuna may have an even worse by-catch of endangered and slow-growing species than purse-seining. There again it may not – the evaluations do not exist. One thing is for sure, it kills a heck or a lot of sharks and turtles. Troll or pole and line-caught tuna are the best alternative.

Prawns/langoustines

The by-catch, generally huge, is the big issue with the wild ones. Trap-caught or responsibly farmed ones are the ethical choice.

Skates and rays

We are told these are long-lived species vulnerable to over-exploitation. The Marine Conservation Society lists them as to be avoided and Marks and Spencer will not stock them for sustainability reasons.

Fish to eat with less conscience

Horse mackerel

Oily fish beloved of Japanese emperors. Usually finds its way into fish meal. Why not eat it instead of farmed salmon?

Blue whiting

Small cod-like fish which could perfectly well substitute for cod in the countries which are fond of it, but in the process of being massively over-exploited for fishmeal. Building a market for it as a table fish, which would mean it was fished for more slowly instead of being caught and pulped for fishmeal, could help to keep it from collapse.

Sand eel
Staple of fishmeal plants, perfectly acceptable cooked like whitebait. So why don't we? Same goes for capelin.

Herring
Though up to a third of catches in the North Sea are estimated to be illegal, the population has grown since the 1970s and it is the closest thing Europe has to a sustainable stock. Good for you, too.

Mackerel
Again, far too much illegal fishing and concern about stocks, but a fast-reproducing fish and in a much better state than Europe's whitefish. Handline fishery in the South West is MSC certified, even though the stock gives cause for concern at times.

Lobster
Some stocks are overfished, but potting is selective and there are proposals to 'ranch' lobster to preserve stocks.

Pacific halibut
MSC certified.

Pollock
On its arduous way to becoming MSC certified, overfished in Russia.

Hoki
OK just now, but New Zealand must act fast if the stock continues to decline.

Pacific salmon
Canned or fresh, better conserved than the Atlantic kind.

Mussels
Farmed.

Tilapia
This and other vegetarian fish are the one true hope of the developing world and could be everybody's future. Might as well get used to cooking them now. Over to you, celebrity chefs.

GLOSSARY

CCAMLR Convention on the Conservation of Antarctic Marine Living Resources. Pronounced 'kammelar'. Scientific body set up in 1982 under the Antarctic Treaty to conserve marine life in the Southern Ocean. Takes particular account of the links between species, such as birds, seals and krill – 'the ecosystem approach'. Enforcement leaves much to be desired.

CITES Convention on International Trade in Endangered Species of Wild Flora and Fauna. Intended to ensure trade in wild animals and plants does not threaten their survival. Entered into force 1975. List of protected species includes relatively few fish.

COLTO Coalition of Legal Toothfish Operators. Australia-based coalition which names and shames toothfish poachers in its web-based Rogues Gallery.

Demersal fish Fish living and feeding on the bottom, such as cod or hake. Sometimes called groundfish.

EEZ Exclusive Economic Zone. The EEZ was a key provision of the 1982 UN Convention on the Law of the Sea (UNCLOS), allowing each coastal state exclusive rights over all resources, whether oil and gas, fish, gravel or minerals within 200 miles of its coast.

FAD Fish Aggregation Device. Fishermen noticed that tropical tuna congregated under logs, whales and other large objects. Now they make their own: a FAD is a wooden frame with (optional) wisps of netting hanging from it. This is launched into the ocean currents attached to a radio or satellite beacon that can relay all sorts of information that indicates the likely presence of fish, including sea temperature, back to the fishing vessel.

FAO UN Food and Agriculture Organisation. Pro-development UN organisation which publishes reports on state of world fisheries.

IATTC Inter-American Tropical Tuna Commission. Established 1950. Regional fisheries organisation for the eastern Pacific, based in La Jolla, California.

ICCAT International Commission for the Conservation of Atlantic Tunas. Established 1969, based in Madrid. Known by conservationists as International Conspiracy to Capture All Tunas, because of its indifferent success in conserving fish populations. Some successes lately in forcing pirate vessels off rogue states' registers and with swordfish in North Atlantic.

ICES International Council for the Exploration of the Sea. Founded 1902. The organisation which promotes and coordinates marine research in the North Atlantic. More than 100 years on, it still hasn't worked out the pre-industrial spawning stock of the North Sea cod.

IOTC Indian Ocean Tuna Commission. Established 1993. Regional fisheries body for tuna in Indian Ocean, based in the Seychelles. Does stock assessments and compiles list of authorised fishing vessels but as yet does not apply any active conservation measures.

ITQ Individual Transferrable Quota. Sometimes known as Individual Fishing Quotas (Alaska). Long-term tradeable rights to fish which adherents claim have been highly successful in ending the 'race to fish'.

IUCN International Union for the Conservation of Nature. Global conservation science network, also called the World Conservation Union. Too many names. Publishes Red List of the world's endangered species on the web.

IWC International Whaling Commission. Club of whaling nations established 1946 to manage whale stocks. If fisheries were run by its latest management rules – developed since commercial whaling ended – hardly a cod boat would put to sea.

MSY Maximum Sustainable (or Sustained) Yield. Optimum relationship between fish stock and fishing effort, the pursuit of which has almost always led to the stock being overfished.

NAFO Northwest Atlantic Fisheries Organisation. Established 1979, based Dartmouth Nova Scotia. Regional organisation for waters beyond the continental shelf. Sixteen members. Current shambles: illegal fishing on edge of Grand Banks.

NEAFC North East Atlantic Fisheries Commission. Regional fisheries organisation. Established 1963 but dates back to between the wars. Based in London. Current shambles: blue whiting, the Hatton Bank.

Pelagic fish Fish living in the upper waters of the open sea such as mackerel, sardine or tuna.

UNCLOS UN Convention on the Law of the Sea. Called for 1967, talked about in 1973, adopted 1982, still not ratified by United States.

Whitefish Fish with white flesh, where the main reserves of fat are in the liver, such as cod, haddock or whiting. As opposed to oily fish such as herring, sprat, mackerel or shellfish.

WWF Formerly World Wildlife Fund, then the World Wide Fund for Nature, now annoyingly just initials.

BIBLIOGRAPHY

*These books provide a good introduction for those coming to the subject of overfishing for the first time. The list of further reading offers more suggestions for the general reader.

Chapter 1
Prime Minister's Strategy Unit, *Net Benefits: A Sustainable and Profitable Future for UK Fishing* (Cabinet Office, March 2004 or www.strategy.gov.uk)

Chapter 2
Carl Safina, *Song for the Blue Ocean** (Henry Holt, 1997)
James R. McGoodwin, *Crisis in the World's Fisheries** (Stanford University Press, 1990)
D. Pauly, V. Christensen, S. Guénette, T. J. Pitcher, U. R. Sumaila, C. J. Walters, R. Watson & D. Zeller, 'Toward sustainability in world fisheries', *Nature*, issue 418, pp 689-95 (2002)
R. Watson & D. Pauly, 'Systematic distortions in world fisheries catch trends', *Nature*, issue 414, pp 534-6 (2001)
Daniel Pauly, 'Why the international community needs to help create marine reserves', Fourth meeting of the UN Open-ended Informal Consultative Process on Oceans and the Law of the Sea, New York, 4 June 2003 (www.un.org)
Ransom Myers & Boris Worm, 'Rapid worldwide depletion of predatory fish communities', *Nature*, issue 423, pp 280-3 (15 May 2003)

Chapter 3
Institute for European Environmental Policy, 'Fisheries agreements with third countries – is the EU moving towards sustainable development?' (www.ieep.org.uk, 2002)
Environmental Justice Foundation, *Squandering the Sea: How Shrimp Trawling is*

Threatening Ecological Integrity and Food Security around the World (EJF, London, 2003)

Chapter 4

O. T. Olsen, *The Piscatorial Atlas of the North Sea, English and St George's Channels* (Taylor & Francis, London, 1883)

Daniel Pauly & Jay MacLean, *In a Perfect Ocean: The State of Fisheries and Ecosystems in the North Atlantic Ocean** (Island Press, 2002)

International Council for the Exploration of the Sea, 'Environmental Status of the European Seas', a quality status report prepared for the German Federal Ministry for the Environment, 2003.

Ole Lindquist, 'The North Atlantic grey whale (*Escherichtius robustus*): An historical account based on Danish-Icelandic, English and Swedish sources dating from *c.* 1000 to 1792' (Universities of St Andrews and Stirling, Scotland, March 2000)

Paul Naylor, *Great British Marine Animals* (Sound Diving Publications, 2003)

J. Nichols, T. Huntington, P. Winterbottom & A. Houg, 'Certification report for the Pelagic Freezer Trawler Association, North Sea Herring Fishery' (Moody Marine, November 2003, or www.msc.org)

S. J. de Groot & H. J. Lindeboom, eds, *Environmental Impact of Bottom Gears on Benthic Fauna in Relation to Natural Resources Management and Protection of the North Sea* (Netherlands Institute for Sea Research, 1994)

Michael Wigan, *The Last of the Hunter Gatherers: Fisheries Crisis at Sea** (Swan Hill Press, 1998)

Chapter 6

Phil Aikman, 'Is deep water a dead end? A policy review of the gold rush for "ancient deepwater" fish in the Atlantic Frontier' (Greenpeace, 1997)

John D. M. Gordon, 'Rockall Plateau now in international waters. The fleets move in', Scottish Association for Marine Science newsletter, no. 22 (October 2000)

John D. M. Gordon, 'The Rockall Trough, Northeast Atlantic: An account of the change from one of the best-studied deep-water ecosystems to one that is being subjected to unsustainable fishing activity', North Atlantic Fisheries Organization, Scientific Council Report N4489 (September 2001)

John D. M. Gordon, 'Fish in deep water', Scottish Association for Marine Science newsletter, no. 27 (2003)

M. Lack, K. Short & A. Willock, 'Managing risk and uncertainty in deep-sea fisheries: lessons from orange roughy', *Traffic*, WWF (2003)

Chapter 7

A. J. Lee, *The Directorate of Fisheries Research, Its Origins and Development* (Ministry of Agriculture, Fisheries and Food, 1992)

Michael Graham, *The Fish Gate** (Faber & Faber, 1943)

Thomas Henry Huxley, 'Address to the International Fisheries Exhibition' (London, 1883)

Helen M. Rozwadowski, *The Sea Knows No Boundaries: A Century of Marine Science under ICES* (International Council for the Exploration of the Sea, 2002)

John Dyson, *Business in Great Waters: The Story of British Fishermen* (Angus & Robertson, 1977)

Ray Beverton & Sidney Holt, *On the Dynamics of Exploited Fish Populations* (HMSO, 1957)

Alan Christopher Finlayson, *Fishing for Truth* (Institute of Social and Economic Research, Memorial University, Newfoundland, 1994)

Carl Walters & Jean-Jacques Maguire, 'Lessons for stock assessment from the northern cod collapse', *Fish Biology and Fisheries*, issue 6, pp 125-37 (1996)

Debora MacKenzie, 'The cod that disappeared', *New Scientist* (16 September 1995)

Chapter 8

Jake C. Rice, Peter A. Shelton, Denis Rivard, Ghislain A. Chouinard & Alain Fréchet, 'Recovering Canadian Atlantic cod stocks: the shape of things to come' (Paper given at a conference organized by International Council for the Exploration of the Sea: The Scope and Effectiveness of Stock Recovery Plans in Fishery Management, Tallinn, Estonia, 2003)

Gareth Porter, *Estimating Overcapacity in the Global Fishing Fleet* (WWF, 1998)

WWF/Spain, 'WTO and fishing subsidies: Spanish case' (Summary of Spanish report in English, WWF, 2003)

Chapter 9

David J. Doulman, 'Global overview of illegal, unreported and unregulated fishing and its impacts on national and regional efforts to sustainably manage fisheries: The rationale for the conclusion of the 2001 FAO international plan of action to prevent, deter and eliminate illegal, unreported and unregulated fishing' (FAO, August 2003)

Hélène Bours (Greenpeace International), 'The Tragedy of Pirate Fishing', International Conference against Illegal, Unreported and Unregulated Fishing, Santiago de Compostela, Spain, 25-26 November 2002

Michael Earle (Fisheries Adviser, Green/EFA Group in the European Parliament), 'The European Union, subsidies and fleet capacity, 1983-2002', UNEP Workshop on the Impacts of Trade-Related Policies on Fisheries and Measures Required for Sustainable Development, 15 March 2002

Justinian, *The Institutes of Justinian, Book Two.*

Garrett Hardin, 'The tragedy of the commons', *Science*, issue 13, vol. 162, pp 1243-8 (December 1968) and *The Concise Encyclopedia of Economics* (Library of Economics and Liberty, www.econolib.org, 2002)

M. Lack & G. Sant, 'Patagonian toothfish: are conservation and trade measures working?', *Traffic*, issue 1, vol. 19 (2001)

Chapter 10

Juan L. Suarez de Vivero & Juan C. Rodriguez Mateos, 'Spain and the sea. The decline of an ideology, crisis in the maritime sector and the challenges of globalisation', *Marine Policy*, issue 26 (2002)

Chapter 11

Nobuyuki Matsuhisa, *Nobu: The Cookbook* (Quadrille, 2001)

Seafood Watch Program (see Monterey Bay Aquarium in Websites below)

Bernadette Clarke, *Good Fish Guide** (Marine Conservation Society, 2002)

Gordon Ramsay with Roz Denny, *Passion for Seafood* (Conran Octopus, 1999)

Environmental Justice Foundation, 'Squandering the seas: how shrimp trawling is threatening ecological integrity and food security around the world' (2003)

E. V. Romanov, 'By-catch in the purse-seine tuna fisheries in the western Indian Ocean', Seventh Expert Consultation on Indian Ocean Tunas, Victoria, Seychelles, 9-14 November 1998

X. Mina, I. Artetxe & H. Arrizabalaga, 'Updated analysis of observers' data available from the 1998-99 moratorium in the Indian Ocean', IOTC proceedings no. 5, pp 340-5 (2002)

Ziro Suzuki (National Institute of Far Seas Fisheries), 'Memorandum on regulatory measures for purse-seine fisheries in the western tropical Indian Ocean', IOTC proceedings no. 5, pp 176-7 (2002)

James Joseph, 'Managing fishing capacity of the world tuna fleet', *Fisheries Circular*, no. 982 (FAO, 2003)

Chapter 12

Stephen Cunningham & Jean-Jacques Maguire, 'Factors of unsustainability and over-exploitation in fisheries' (FAO, February 2002)

Report and documentation of the International Workshop on Factors of Unsustainability and Overexploitation in Fisheries, Bangkok, Thailand, 4-8 February 2002 (FAO, 2002)

Jake Rice, 'Sustainable uses of the ocean's living resources', *Isuma*, Canadian Journal of Policy Research (Autumn 2002)

Chapter 13

Donald R. Leal, *Fencing the Fishery: A Primer on Ending the Race for Fish* (Center for Free Market Environmentalism, 502 South 19th Ave, Suite 211, Bozeman, Montana 59718-6827 or www.perc.org)

Laura Jones & Michael Walker, eds, *Fish or Cut Bait; The Case for Individual Transferable Quotas in the Salmon Fishery of British Columbia* (Fraser Institute, 1997)

Chapter 14

F. R. Gell & C. M. Roberts, *The Fishery Effects of Marine Reserves and Fishery Closures* (WWF-US, Washington, 2003)

Bill Ballantine, 'Marine reserves in New Zealand: the development of the concept and the principles', proceedings of an International Workshop on Marine Conservation for the New Millennium, pp 3-38, Korean Ocean Research and Development Institute, Cheju Island, November 1999

Bill Ballantine, '"No take" Marine Reserve Networks Support Fisheries,' paper given at 2nd World Fisheries Congress, Brisbane, 1996.

These and other Ballantine papers can be found on www.marine-reserves.org.nz/pages/ballantine.html or www.2.auckland.ac.nz

Chapter 15

Guardian poll about the British public's ethical purchasing habits published 28 February 2004.

Monterey Bay Aquarium, Seafood Watch (see Websites)

Blue Ocean Institute, Seafood Guide (see Websites)

Marine Stewardship Council papers on pollock, hoki, Thames herring, etc., including those for and against certification, can be seen on www.msc.org

Jim Gilmore, 'The MSC and its significance' (At-Sea Processors Association, October 2003)

Chapter 16

Charles Clover, 'The Price of Fish' *Telegraph* Magazine (26 October 1991)

Stuart M. Barlow, 'The world market overview of fish meal and fish oil', Second Seafood and By-products Conference, Alaska (www.iffo.org.uk, November 2002)

'The promise of a blue revolution', *The Economist* (7 August 2003)

Information on deep-water fisheries and the biological status of the stocks is contained in the Advisory Committee on Fishery Management (ACFM) report (ICES, 12 May 2003)

Ronald A. Hites, Jeffrey A. Foran, David O. Carpenter, M. Coreen Hamilton, Barbara A. Knuth & Steven J. Schwager, 'Global assessment of organic contaminants in farmed salmon', *Science*, vol. 303, p. 226 (2004)

John Humphrys, *The Great Food Gamble* (Hodder and Stoughton, 2001)

Chapter 17

Marion Shoard, *Theft of the Countryside* (MT Smith, 1980)

National Audit Office, 'Fisheries Enforcement in England', a report by the Comptroller and Auditor General (April 2003).

FURTHER READING

Fishing News, a weekly newspaper aimed at fishermen

Fishing News International, a monthly newspaper aimed at fishermen

Sylvia A. Earle, *Sea Change* (Constable, 1995)

Richard Ellis, *The Empty Ocean* (Shearwater, 2003)

Mike Holden, *The Common Fisheries Policy*, with an update by David Garrod (Fishing News Books, 1984)

Mark Kurlansky, *Cod: A Biography of the Fish that Changed the World* (Jonathan Cape, 1997)

Arthur F. McEvoy, *The Fisherman's Problem* (CUP, 1986)

Farley Mowat, *Sea of Slaughter* (Bantam, 1984)

Daniel Pauly, *On the Sex of Fish and the Gender of Scientists* (Chapman & Hall, 1994)

WEBSITES

Blue Ocean Institute, Seafood Guide
www.blueoceaninstitute.org

Rogues' gallery of fishermen
www.colto.org

Food and Agriculture Organization of the UN
www.fao.org

FishBase lists 28,500 species of fish, with information about everything from their rarity and ability to withstand fishing pressure, to how they reproduce.
www.fishbase.org

Stock assessments in the North Atlantic compiled by the International Council for the Exploration of the Sea.
www.ices.org

Monterey Bay Aquarium, Seafood Watch
www.mbayaq.org

List of endangered species produced by the International Union for the Conservation of Nature.
www.redlist.org

INDEX

ACKNOWLEDGEMENTS

I FIND IT difficult at times to believe that I have actually written this book, seven years and two agents after I tried to interest publishers (though not Ebury Press) in such an idea and thirteen years after it first occurred to me that the world needed such a book. It is quite difficult to thank all the people who have helped me to get it published, let alone in the right order. For yes, this is one of those publishing stories about trying and trying and only eventually succeeding.

In 1997, my then agent, Xandra Bingley, who has gone on to take up writing full time, was kind enough to go the rounds with a proposal. We found that publishers thought the subject of what commercial fishing was doing to the world we live in, and the fish we eat, interesting but not a commercial proposition. This meant there was not enough to support me while writing it, so I lined up sponsorship from WWF and Unilever who had then just set up an organisation to certify fisheries that were sustainably managed. This enabled me to take three months off my job at *The Daily Telegraph*, just before the 1997 election to do some research. Then the small publisher, who was all that we had been able to find, went back on his original agreement and the project crashed to the ground. I went back to work and published what I could that was finished on the front page of *The Daily Telegraph* – it makes up part of Chapter 10, The Slime Trail. I explained what had happened to my sponsors, who were kind enough to say, without revealing their disappointment, that my debt had been discharged.

Over the next six years, I buffed up the proposal, tried to persuade the odd agent or publisher to take it on, with no success. Just when I had given up hope of ever interesting anyone in it, or taking enough time off to write it, I got a call from Brendan May, a friend and the chief executive of the Marine Stewardship Council, to say that he had been approached by an agent with a new business who wanted to find someone to write a book about fish. His name was Ivan Mulcahy. Brendan told Ivan I had probably already written it, which was not true, but it meant Ivan made the call that began the story of this book.

So my thanks must begin with Brendan – whose organisation I have therefore felt obliged to give as hard a kicking as I have anyone else's – for his faith in me and for introducing me to Ivan, who persuaded me to write this book, made it possible for me to take the time to do it, and who was responsible for recognising my better ideas, deleting my worst ones, and for reading every word several times over in the hope that I might live up to my original intentions. I owe both a great debt. After them, I must thank Fiona MacIntyre at Ebury Press for taking it on with such confidence and seeing off what, to my surprise, was a great deal more interest in the subject of fish in the publishing business than had existed seven years previously. Thanks after that must go to those who responded to my nervous request to write persuading publishers that this proposal was a project of some merit: I am immensely grateful on this count to HRH the Prince of Wales, Margaret Atwood, Sebastian Faulks, Brendan May and George Monbiot. I must also thank Charles Moore at *The Daily Telegraph* for giving me the time off to write the book, the news desk for putting up with me at difficult moments, and *The Daily Telegraph* generally for sending me on one or two assignments which have ended up being part of it.

Perhaps it was the urgent need to tell this story of what is going on in the sea in the pursuit of what we eat that made so many people so very kind to me. Along the way, recently or not so recently, the following people offered invaluable help, either by being particularly generous with their time and ideas, or being kind enough to read and offer suggestions on my drafts, or both. These were: John Beddington, Brendan May, Euan Dunn, Orri Vigfusson, Ragnar Arnason, Daniel Pauly, Joe Horwood, Mark Tasker, Jonathan Peacey, Sidney Holt, Monica Verbeek, Sue Windebank, Mike Sutton, John Gummer, John Shepherd, Tommy Finn, John Williams, John Pope, Raul Garcia, Georgina Mace, Papa Samba Diouf, Odd Nakken, Alan Simcock, Han Lindeboom, Bill Ballantine,

George Clement, Alastair O'Rielly, Richard Sandbrook, Ken Stump, Michael Earle, Hélène Bours, Geoff Kirkwood and my sister Gillian Clover.

I must thank those who gave up their time and contributed thoughts and encouragement, though often at some remove, and in one or two cases without ever having met me face to face – a new definition of kindness in the information age. These include Jake Rice, Ransom Myers, Neil Fletcher, Sir Tipene O'Regan, Rognavaldur Hannesson, Chris Hutchings, Jenny Hodder, Jim Gilmore, Brian O'Riordan and Thelma Wilson. I must also thank all the people I interviewed for this book. As this is a non-fiction book most of their names are already recorded in it and, to save the reader's patience, I intend not to list them again, but everyone who is mentioned in this book will know how generously, without exception, they gave up their time in the hope that I would emerge better informed. I salute them for expressing their views so memorably, in ways that have become the very fabric of what I have written. If there is anyone who feels they should have been included, and they are not, mine is the journalists' excuse: this book was eventually written very fast, it is late, time presses and my brain is not at its best after writing from morning to late at night. My hope is that, when it comes to reporting what they actually said, or their ideas, I did not let them down.

The last and most important thank-you is to my wife Pamela for putting up with me, supporting me when no one else did, and for reading every word that I wrote to make sure that if I was to make a fool of myself, at least it was deliberate. My love and thanks to her.

Charles Clover,
Dedham, 2004